SAM1R06c

Decision Analysis/ Implementation Guidance

STRATEGIC ASSET MANAGEMENT AND COMMUNICATION

by:

Duncan Rose
GHD Consulting Inc.

2010

WERF

The Water Environment Research Foundation, a not-for-profit organization, funds and manages water quality research for its subscribers through a diverse public-private partnership between municipal utilities, corporations, academia, industry, and the federal government. WERF subscribers include municipal and regional water and wastewater utilities, industrial corporations, environmental engineering firms, and others that share a commitment to cost-effective water quality solutions. WERF is dedicated to advancing science and technology addressing water quality issues as they impact water resources, the atmosphere, the lands, and quality of life.

For more information, contact:
Water Environment Research Foundation
635 Slaters Lane, Suite 300
Alexandria, VA 22314-1177
Tel: (703) 684-2470
Fax: (703) 299-0742
www.werf.org
werf@werf.org

This report was co-published by the following organization.

IWA Publishing
Alliance House, 12 Caxton Street
London SW1H 0QS, United Kingdom
Tel: +44 (0) 20 7654 5500
Fax: +44 (0) 20 7654 5555
www.iwapublishing.com
publications@iwap.co.uk

Library of Congress Catalog Card Number: 2009928929

IWAP ISBN: 978-1-84339-332-0/1-84339-332-8

This report was prepared by the organization(s) named below as an account of work sponsored by the Water Environment Research Foundation (WERF). Neither WERF, members of WERF, the organization(s) named below, nor any person acting on their behalf: (a) makes any warranty, express or implied, with respect to the use of any information, apparatus, method, or process disclosed in this report or that such use may not infringe on privately owned rights; or (b) assumes any liabilities with respect to the use of, or for damages resulting from the use of, any information, apparatus, method, or process disclosed in this report.

GHD Consulting, Inc.

The research on which this report is based was developed, in part, by the United States Environmental Protection Agency (EPA) through Cooperative Agreement No. CR-83155901 with the Water Environment Research Foundation (WERF). However, the views expressed in this document are not necessarily those of the EPA and EPA does not endorse any products or commercial services mentioned in this publication. This report is a publication of WERF, not EPA. Funds awarded under the Cooperative Agreement cited above were not used for editorial services, reproduction, printing, or distribution.

This document was reviewed by a panel of independent experts selected by WERF. Mention of trade names or commercial products or services does not constitute endorsement or recommendations for use. Similarly, omission of products or trade names indicates nothing concerning WERF's or EPA's positions regarding product effectiveness or applicability.

ACKNOWLEDGMENTS

The researchers would like to acknowledge the kind assistance of all those who contributed to this project.

Research Team

Principal Investigator:
Duncan Rose
GHD Consulting, Inc.

Issue Area Team Members

Kevin Hadden
Orange County Sanitation District, CA

Greg Kane
Sydney Water, Australia

Wayne Green
Green Management

Ken Jacob
Cobb County Water System, GA

John Colbert
Massachusetts Water Resources Authority, MA

Terry Martin
Seattle Public Utilities, WA

Steve Krai
Los Angeles Sanitation District, CA

Adam Gronski
Metropolitan Water Reclamation District, Chicago, IL

Working Party Members

Elizabeth Kelly
Seattle Public Utilities, WA

Thais Vitigliano
Washington Suburban Sanitary Commission, Washington, D.C.

John O'Neill
Aaron Witt
Stacy Gallick
Johnson County Wastewater, MO

Andre Blaylock
Jim Hewitt
City of Akron, OH

John Sullivan
Boston Water and Sewer Commission, MA

Ed Jones
County of Fairfax, VA

Jeff Brochtrup
Mike Simon
Madison Metropolitan Sewer District, WI

Maura Bonnerans
East Bay Municipal Utilities District

Guy Graham
City of Gresham, OR

Joel Thompson
Fairfax Water

Water Environment Research Foundation Staff

Director of Research: Daniel M. Woltering, Ph.D.
Program Director: Walter L. Graf, Jr.

ABSTRACT AND BENEFITS

Abstract:

Strategic Asset Management (SAM) is a structured framework that assists asset managers to make cost-effective investment decisions for existing and new assets to sustain long-term performance and levels of service. Track 3 of WERF's Strategic Asset Management Project was developed to "provide guidance on implementing SAM…and develop analytic tools for SAM implementation" – the Decision Support Tools and Implementation Guidance goal. The Decision Support Tools and Implementation Guidance goal was organized around two major elements:

- **Analytic/Decision Support Tools** – Develop staged, user-friendly analytic and decision support tools (gap analysis, cost benefit analysis, risk management, life cycle costing, condition assessment selection tool, etc.) for SAM implementation and continuous improvement.
- **Implementation Tool** – Provide practical, accessible guidance for an implementation tool to undertake SAM in incremental stages leading to a well-planned, structured, and progressive implementation to suit the needs of a diverse range of utilities' communication practices.

The present research includes results of a web-based search for existing asset management tools and guidance materials, a prioritized list of tools that need to be developed, and critical factors that contribute to, or inhibit, successful implementation of SAM. These activities resulted in part from discussions of a Practitioner Tool-Review Working Group that included both SAM "starters" and experienced "veterans." The report also refines the SAM implementation pathways in SIMPLE, and provides examples of pathway selection based on organization characteristics and cost and performance factors. A structured process is presented for identifying information needs to support SAM implementation. Project materials will be added to SIMPLE, WERF's web-based asset management knowledge base.

Benefits:

Asset management can be defined as a way of management thinking and a set of best practices built around a structured framework that assist asset managers in making the most cost effective investment decisions in existing and new assets to sustain long term performance of the assets in an environment of limited resources where these decisions represent an integration of operations, maintenance, and capital intervention strategies. At the heart of the framework is the concept of business risk – a metric that measures the interaction of the probability of failure, the consequence of failure, and risk mitigation strategies. This research:

- Facilitates the adoption of best practice in asset management in the public water utility industry.
- Augments information in SIMPLE, an asset management knowledge base that is a structured repository of a broad range and deeply substantive collection of materials oriented toward the practicing asset manager.
- Addresses the why and how of asset management at levels for novice, intermediate, and advanced practitioners.

Keywords: Asset management, risk, business risk exposure, critical success factors, pathways to implementation, asset management tools, gap analysis, best practice.

TABLE OF CONTENTS

LIST OF TABLES

LIST OF FIGURES

LIST OF ACRONYMS

ASTM	American Society of Testing and Materials
AWWA	American Water Works Association
BCA	Benefit Cost Analysis
BRE	Business Risk Exposure
CAP	Condition Assessment Protocol
CIP	Capital Improvement Program
CLR	Confidence Level Rating
CSIRO	Commonwealth Scientific and Industrial Research Organization (Australia)
CSO	Combined Sewer Overflow
CUPSS	Check Up Program for Small Systems
FEMA	U.S. Federal Emergency Management Association
FHWA	Federal Highway Association
LOS	Level of Service
MTBF	Mean Time Between Failure
MHW	Montgomery Watson Harza
NACWA	National Association of Clean Water Agencies
NZWWA	New Zealand Water and Wastewater Association
O&M	Operations and Maintenance
PdM	Predictive Maintenance
RCM	Reliability Centered Maintenance
SAM	Strategic Asset Management
SIMPLE	Sustainability Infrastructure Management Program Learning Environmental
SSO	Sanitary Sewer Overflow
TEAMS	Total Electronic Asset Management System
U.S. EPA	U.S. Environmental Protection Agency
USDOT	U.S. Department of Transportation
UKWIR	United Kingdom Water Industry Research
VSAT	Vulnerability Self-Assessment Tool
WERF	Water Environment Research Foundation
WRC	Water Research Council
WRF	Water Research Foundation (formerly AwwaRF)

EXECUTIVE SUMMARY

Utilities throughout the world are faced with the challenge of managing their water and wastewater assets to meet stakeholder and customer levels of service with limited funds. In the United States, the U.S. Environmental Protection Agency released a study in 2002 that highlighted the funding gap between actual and needed investment in water and wastewater infrastructure, estimated at $122 billion for wastewater and $102 billion for drinking water under current revenue scenarios. The projected funding gap is a result of aging infrastructure and increasing population and quality demands on water systems. Closing the gap likely will require both increases in funding and more efficient management of water and wastewater assets, e.g., through application of Strategic Asset Management (SAM) approaches.

In response to the needs of its utility members, the Water Environment Research Foundation initiated a research program in 2006 to develop guidance and decision support tools for communication and implementation of SAM. The research effort is composed of four separate, but inter-related tracks:

Track 1: Public Communication to build support for SAM implementation

Track 2: Benchmarking and Case Studies of SAM programs

Track 3: Decision Analysis and Implementation Guidance for SAM

Track 4: Remaining Asset Life estimation techniques

The materials presented in this report were developed in support of Track 3, to catalog available SAM tools, identify and prioritize a core set of needed tools, explore factors that contribute to successful asset management, refine pathways for implementing SAM, and develop the information needed to support SAM implementation. Project results described in this report will be added to SIMPLE, WERF's web-based asset management knowledge base.

Chapter 2.0 provides an inventory of non-commercial asset management tools that were available in the public domain at no or very little cost at the time of the search. Very few tools were identified that were web-based, interactive, and applicable to the water utility community. Where possible, the authors also make special note of methodological documents representing best practice that were in the public domain.

Chapter 3.0 presents a 10-step process model for building an asset management plan and identifies tools that assist each step of the process. Over 20 tools are described and prioritized for development based on recommendations of a working group of practitioners from a workshop held in January 2008. The recommended tools include hierarchical asset registries, condition assessment, residual life, life cycle costing, and level of service tools, as well as tools to assess business risk, benefit/cost, and funding optimization.

Chapter 4.0 presents the factors that practitioners believe contribute to successful implementation and maintenance of asset management programs, or were inhibitors to program success. Recurring themes in workshops held with practitioners and novice "beginners" include the importance of management commitment, adequate staff and funding, clearly defined benefits for stakeholders, change management within the organization, phased implementation, and information feedback for continuous improvement. Workshop participants also noted the importance of building a business case for investment in water infrastructure, including efforts to minimize asset life cycle costs and to leverage "whole of government" savings by adopting best asset management practices in all sectors that manage physical assets.

Chapter 5.0 describes seven pathways for implementing asset management, ranging from approaches that focus on single problematic assets to an asset management plan for an entire asset portfolio. The pathways are refinements of the original "Six Alternative Routes to Implementation" found in the first version of SIMPLE - Sustainable Infrastructure Management Program Learning Environment. The chapter presents a checklist of organizational factors and asset cost and performance factors that suggest suitability of particular implementation pathways, and illustrates the pathway selection process using 3 hypothetical examples.

Chapter 6.0 presents a framework for determining the information requirements for SAM decision-making. A core set of information elements is described to allow asset managers to identify the current state of assets, monitor levels of service to meet customer expectations, identify high risk critical assets, optimize investment strategies, and determine long-term funding needs.

CHAPTER 1.0

INTRODUCTION

1.1 Strategic Asset Management (SAM) Contract Team

GHD forms part of a consortium that is conducting research for the Water Environment Research Federation (WERF) on Strategic Asset Management, or SAM (WERF Project SAM1R06). The consortium is comprised of a prime consultant, EMA Inc., and a group of other partner firms and organizations including:

- GHD Consulting Inc.
- CSIRO – Australia
- Montgomery Watson Harza (MWH)
- Virginia Tech's School of Construction and Asset Management

Each of the partners had a responsibility to execute certain elements of the overall WERF contract.

1.2 Overall SAM Challenge Scope

Utilities throughout the world are faced with the challenge of managing their water and wastewater assets to meet stakeholder and customer levels of service with limited funds. In the United States, the U.S. Environmental Protection Agency released a study in 2002 that highlighted the funding gap between actual and needed investment in water and wastewater infrastructure[1], estimated at $122 billion for wastewater and $102 billion for drinking water under current revenue scenarios. The projected funding gap is a result of aging infrastructure and increasing population and quality demands on water systems. Closing the gap likely will require both increases in funding and more efficient management of water and wastewater assets, e.g., through application of Strategic Asset Management (SAM) approaches.

In response to the needs of its utility members, the Water Environment Research Foundation initiated a research program in 2006 to develop guidance and decision support tools for communication and implementation of SAM. The research effort is composed of four separate, but inter-related tracks:

Track 1: Public Communication to build support for SAM implementation

Track 2: Benchmarking and Case Studies of SAM programs

Track 3: Decision Analysis and Implementation Guidance for SAM

Track 4: Remaining Asset Life estimation techniques

[1] U.S. Environmental Protection Agency. The Clean Water and Drinking Water Infrastructure Gap Analysis. Washington, D.C. 2002.

One of the main goals and objectives of the SAM Project was characterized as follows:

> "1. Develop a communication package (advocacy/awareness tools) for key stakeholders (public officials and policy makers, utility managers and rate payers) for SAM adoption; provide guidance on implementing SAM in wastewater utilities (small, medium and large sizes) in a structured, step-by-step, and progressive manner consistent with available resources and a long-term strategy; and develop analytic tools for SAM implementation – gap analysis, cost –benefit analysis for making sound asset management decisions, risk management tools, condition assessment selection tool etc."

1.3 Decision Support Tools and Implementation Guidance

Track 3 of the Project was developed to "provide guidance on implementing SAM…and develop analytic tools for SAM implementation" – the Decision Support Tools and Implementation Guidance goal. In Track 3, the Decision Support Tools and Implementation Guidance goal was organized around two major elements:

Analytic/Decision Support Tools – Develop staged, user-friendly analytic and decision support tools (gap analysis, cost benefit analysis, risk management, life cycle costing, condition assessment selection tool, etc.) for SAM implementation and continuous improvement.

Implementation Tool – Provide practical, accessible guidance for an implementation tool to undertake SAM in incremental stages leading to a well-planned, structured, and progressive implementation to suit the needs of a diverse range of utilities' communication practices.

Task Order 2 provided for the following

- Set up Practitioner Tool-Review Working Group and Establish Framework
- Catalog Available "Commercial Off-The-Shelf" SAM tools
- Identify and Prioritize Core Set of Tools
- Identify Critical Success Factors for Implementing SAM
- Review and Modify the Gap Analysis Tool
- Refine the "Six Alternative Routes to Implementation" Tool
- Develop Information Requirements for SAM Decision Making

The materials presented in the following sections were developed as a result of executing Task Order 2 to achieve the Project goal. Relevant developed materials are in the process of being added to SIMPLE, WERF's web based asset management knowledge base.

CHAPTER 2.0

CATALOG OF EXISTING DECISION SUPPORT TOOLS

The research team conducted an international search to identify available non-commercial decision support tools. An objective of the challenge was to identify any decision support tools that can be made available to WERF/Water Research Foundation members without charge and that can be adapted into the asset management framework around which the SAM project is organized.

2.1 Objectives

Simply because a tool has been developed somewhere in the world and there is no charge for its usage, does not necessarily mean it will be suitable for WERF and Water Research Foundation – most of whom are U.S.-based utilities. Consequently once the tools have been identified some evaluation has to take place to determine whether they might be suitable for use by WERF and Water Research Foundation members. Once any such tools have been identified there will be a range of decision support areas for which there are no available non-commercial tools. The Project Team has identified a prioritized list of tools that need to be developed and customized for WERF and Water Research Foundation members (see separate report entitled "WERF SAM Tool Prioritization").

The sections which follow set out the results of an extensive online international search for asset management decision support tools.

2.2 Methodology

An online search was conducted of websites of many different global associations and governmental agencies to identify decision support tools. In addition searches were undertaken using certain key words and phrases relevant to decision making support, including:

- Life cycle costing
- Infrastructure assets
- Asset valuation
- Asset renewals
- Business risk exposure
- Risk
- Asset Management tools

In many cases, the search led to resource links which described written materials that are available through a government agency. These resources typically consist of a mixture of books, publication, guides, CD-ROMs, and in a few cases actual tools.

Because of its historical pre-eminence of the region in asset management practices and processes, a number of Australian/New Zealand based websites were also searched. These include the websites of water associations in Australia/New Zealand.

In the context of this project, "tools" are considered to be decision tools, process guides, or standards for management of public assets. To ensure the research was as complete as possible, a number of publications were identified in the initial scan that later were determined not to be suitable for inclusion in the project. On the other hand, a number of publications and guidelines were suitable, but a charge is associated with their use; given their relevance, these tools have been included in this report even though there is a charge associated with them.

In many cases there would be some free material and some commercial material available within the same "publications" link on a website. Again where a tool or guide was identified, but it is commercial, it has been noted in this report for the sake of completeness.

2.3 Search Results

The tools identified, and their associated websites, are summarized in Table 2-1 and in Appendix A. Very few tools were identified that were web-based, interactive, and applicable to the water utility community.

However, a number of useful guides and decision support tools have been developed by such governmental agencies as Treasury Departments and by agencies responsible for managing road assets. While many of the fundamental concepts associated with asset management in government departments or at roads management agencies are similar to water utilities, the guides are not completely transferable to the water utility industry. For example there are many pavement management tools available to assist in the management of an important asset in roads context – pavements. However, WERF and Water Research Foundation members have a much better understanding of water assets than roads assets so the guides or tools are not necessarily transferable from one sector to the other.

Even though the tools identified may not be suitable for this project, the structure of the tools/guides etc. developed in other contexts may provide a very useful basis for developing tools specific to the needs of the water industry as the principles upon which asset management decision making are based, transcend asset type.

2.4 Summary of Asset Management Tools

The tools identified in the web search are summarized in Table 2-1, categorized by their applicability to various sectors of asset management – life cycle costing, valuation, etc. A number of the tools have some relevance to several sectors of asset management, and thus appear in multiple categories in the table. Each table entry also is coded as a tool (T) or guide (G), and as non-commercial (N) (i.e., free of charge) or commercial (C) (i.e., available for a fee).

Table 2-1. Catalog of Existing Tools and Guides for Asset Management.

Asset Management Element	Tool Name	Description[a]	Source
Asset Registry	CUPSS – Check Up Program for Small Systems	applicable to water systems with less than 1000 connections or serving up to 3300 persons (T,N)	U.S. EPA
	TEAMS – Total Electronic Asset Management System	AM software for small to medium wastewater utilities (T,N)	U.S. EPA
	Asset Management - A Handbook for Small Water Systems	inventory assets and develop a long-term maintenance plan (G,N)	U.S. EPA
	Taking Stock of your Water System - A Simple Inventory for Very Small Drinking Water Systems	Additional information on asset management and planning and inventory worksheets (G,N)	U.S. EPA
	Strategic Planning: A Handbook for Small Water Systems	Overview of the planning process and worksheets to use in developing a strategic plan (G,N)	U.S. EPA
	SIMPLE – Sustainable Infrastructure Management Program Learning Environment	Web-based knowledge base provides state-of-the-art training, guidance, research, and tools for AM practitioners and for utilities of all types, sizes, and levels of AM. (T,C)	WERF
	PHAST TM	Allows characterization of military, federal, and commercial assets supporting the military health system (G,C)	LMI Gov. Consult.
	International Infrastructure Management Manual	Starts with basic asset management principles to practical steps for implementing advanced asset management systems (G,C)	INGENIUM
Condition Assessment	Mitigation BCA (Benefit Cost Analysis) Toolkit	CD-ROM developed for use in assessment of disasters which has some relevance to the water industry (T,N)	FEMA
	Standardized System to Assess the Condition of Infrastructure in Edmonton	based on five-point scale across three failure criteria – level of service, physical mortality, and capacity (T,N)	City of Edmonton, CA
	TEAMS – Total Electronic Asset Management System	AM software for small to medium wastewater utilities (T,N)	U.S. EPA
	Vulnerability Self-Assessment Tool (VSAT)	National Association of Clean Water Agencies (NACWA) software was developed to support water and wastewater utility vulnerability assessments using a qualitative risk assessment methodology. (G,N)	NACWA
	Emergency Response Guidelines for Wastewater Systems	Assists wastewater systems in the preparation of Emergency Response Plans (G,N)	NACWA
	Asset Based Vulnerability Checklist for Wastewater Utilities	Help identify and evaluate a wide range of vulnerabilities that could place assets in jeopardy (G,N)	NACWA
	GRASP	An index based risk assessment tool for mapping changes in level of risk exposure along a pipeline (T,C)	WRc
	SIMPLE – Sustainable Infrastructure Management Program Learning Environment	Web-based knowledge base provides state-of-the-art training, guidance, research, and tools for AM practitioners at utilities of all types, sizes, and levels of AM. (T,C)	WERF

Asset Management Element	Tool Name	Description[a]	Source
	Condition Assessment Strategies and Protocols for Water and Wastewater Utility Assets	Research report. Project 03-CTS-20CO (G,C)	WERF
	New Zealand Pipe Inspection Manual -2006 3rd Ed	Updated inspection manual with expanded CCTV section (G,C)	WaterNZ
	New Zealand Infrastructure Grading Guidelines – 1999	Condition grading of infrastructure assets (G,C)	WaterNZ
	International Infrastructure Management Manual	Starts with basic asset management principles to practical steps for implementing advanced asset management systems (G,C)	INGENIUM
	Assessment of Non-Uniform Corrosion in Copper Piping	Assesses the prevalence of non uniform corrosion in copper piping within the North American drinking water industry. (G,C)	Water RF
	Long Term Performance Prediction for PVC Pipes	Reviews existing and proposed accelerated material life testing methods that can be applied to buried PVC pipes. (G,C)	Water RF, CSIRO, UKWIR
	Failure of Pre stressed Concrete Cylindrical Pipes	Develops a general evaluation matrix to help utilities identify pre-stressed concrete cylinder pipe (PCCP) with the highest risk of failure in their systems (G,C)	Water RF U.S. EPA
	Workshop on Condition Assessment Inspection Devices for Water Transmission Mains	Conducts a state-of-the-art literature review of non-interruptive condition assessment inspection devices for large diameter transmission mains (greater than 12 inches). (G,C)	Water RF
	Long Term Performance Prediction for PE Pipes	Evaluates and recommends accelerated material life testing methods for buried PE pipes. (G,C)	Water RF, CSIRO, UKWIR
Residual Life	TEAMS – Total Electronic Asset Management System	AM software for small to medium wastewater utilities (T,N)	U.S. EPA
	SIMPLE – Sustainable Infrastructure Management Program Learning Environment	Web-based knowledge base provides state-of-the-art training, guidance, research, and tools for AM practitioners at utilities of all types, sizes, and levels of AM. (T,C)	WERF
	International Infrastructure Management Manual	Starts with basic asset management principles to practical steps for implementing advanced asset management systems (G,C)	INGENIUM
Asset Valuation	CUPSS – Check Up Program for Small Systems	applicable to water systems with less than 1000 connections or serving up to 3300 persons (T,N)	U.S. EPA
	TEAMS – Total Electronic Asset Management System	AM software for small to medium wastewater utilities (T,N)	U.S. EPA
	Building Life Cycle Costing (BLCC) Program	Software to provide computational support for the analysis of capital investments in buildings (T,N)	US DOE
	ECONPAK	Software for economic evaluation of projects (T,N)	US ACOE
	Valuation of Road Infrastructure in Australia/New Zealand	(G,N)	Austroads
	Economic Evaluation of Road Investment Proposals	Economic valuation of road investment (G,N)	Austroads
	Valuing the Environment and Other Externalities	(G,N)	Austroads
	Valuation Guidelines for Plant, Property and Equipment	Guidance on asset valuation (G,N)	NZ Dept of

Asset Management Element	Tool Name	Description[a]	Source
			Treasury
	Fair Value Asset Valuation	Provides advice on methodologies for obtaining appropriate and consistent valuation of non-current physical assets for effective asset management and for financial reporting. (G,N)	Dept of Sustainability Australia
	SIMPLE – Sustainable Infrastructure Management Program Learning Environment	Web-based knowledge base provides state-of-the-art training, guidance, research, and tools for AM practitioners at utilities of all types, sizes, and levels of AM. (T,C)	WERF
	International Infrastructure Management Manual	Starts with basic asset management principles to practical steps for implementing advanced asset management systems (G,C)	INGENIUM
Replacement Costing	Building Life Cycle Costing (BLCC) Program	Software to provide computational support for the analysis of capital investments in buildings (T,N)	US DOE
	The Life Cycle Cost Analysis Primer	Office of Asset Management life cycle costing guidelines. (T,N)	USDOT, FHWA
	Valuation Guidelines for Plant, Property and Equipment	Guidance on asset valuation (T,N)	NZ Dept of Treasury
	ECONPAK	Software for economic evaluation of projects (T,N)	US ACOE
	Waterfowl	life cycle costing (T,C)	WRc
	Mitigation BCA (benefit cost analysis) Toolkit	a CD-ROM developed for use in assessment of disasters which has some relevance to the water industry (T,C)	FEMA
	SIMPLE – Sustainable Infrastructure Management Program Learning Environment	Web-based knowledge base provides state-of-the-art training, guidance, research, and tools for AM practitioners at utilities of all types, sizes, and levels of AM. (T,C)	WERF
	International Infrastructure Management Manual	Starts with basic asset management principles to practical steps for implementing advanced asset management systems (G,C)	INGENIUM
	Standard practice for least cost (life cycle) analysis of concrete culvert, storm sewer, and sanitary sewer systems and other buried conduits.	procedure for using life cycle cost (LCC) analysis techniques (G,C)	ASTM
	Standard practice for measuring life cycle cost analysis of plastic pipe used in culverts, storm sewers and other buried conduits	procedure for using life cycle cost (LCC) analysis techniques (G,C)	ASTM
	Standard practice for life cycle cost analysis of corrugated metal pipe used for culverts, storm sewers, and other buried conduits	procedure for using life-cycle cost (LCC) analysis techniques to evaluate alternative drainage system designs using corrugated metal pipe that satisfies the same functional requirements. (G,C)	ASTM
Level of Service	LOSPLAN	a software application that computes Levels of Service on a five point scale given a series of usage scenarios (T,N)	Florida State DOT
	Guidelines for development of Levels of Service Framework based on Community Consultation	(G,N)	Austroads

Asset Management Element	Tool Name	Description[a]	Source
	SIMPLE – Sustainable Infrastructure Management Program Learning Environment	Web-based knowledge base provides state-of-the-art training, guidance, research, and tools for AM practitioners at utilities of all types, sizes, and levels of AM. (T,C)	WERF
	International Infrastructure Management Manual	Starts with basic asset management principles to practical steps for implementing advanced asset management systems (G,C)	INGENIUM
Business Risk	Mitigation BCA (Benefit Cost Analysis Toolkit	CD-ROM developed for use in assessment of disasters which has some relevance to the water industry (T,N)	FEMA
	Standardized System to Assess the Condition of Infrastructure in Edmonton	based on five-point scale across three failure criteria – level of service, physical mortality, and capacity (T,N)	City of Edmonton, CA
	Vulnerability Self-Assessment Tool (VSAT)	software developed to support water and wastewater utility vulnerability assessments using a qualitative risk assessment methodology. (G,N)	NACWA
	Emergency Response Guidelines for Wastewater Systems	Assists wastewater systems in the preparation of Emergency Response Plans (G,N)	NACWA
	Asset Based Vulnerability Checklist for Wastewater Utilities	Help identify and evaluate a wide range of vulnerabilities that could place assets in jeopardy (G,N)	NACWA
	Information Security Risk Assessment	This simple checklist provides a high-level view of common security practices (G,N)	California Office of Information Security and Privacy
	Performance Risk Assessment	Risk-based method and assessment tool that analyzes the probability and consequence of failure (T,C)	LMI Govt Consulting
	SIMPLE – Sustainable Infrastructure Management Program Learning Environment	Web-based knowledge base provides state-of-the-art training, guidance, research, and tools for AM practitioners at utilities of all types, sizes, and levels of AM. (T,C)	WERF
	International Infrastructure Management Manual	Starts with basic asset management principles to practical steps for implementing advanced asset management systems (G,C)	INGENIUM
Capital/ Maintenance Investment Optimization	Mitigation BCA (Benefit Cost Analysis) Toolkit	CD-ROM developed for use in assessment of disasters which has some relevance to the water industry (T,N)	FEMA
	TEAMS – Total Electronic Asset Management System	AM software for small to medium wastewater utilities (T,N)	U.S. EPA
	ECONPAK	Software for economic evaluation of projects (T,N)	US ACOE
	Cost Benefit Analysis Primer	An overview of Cost Benefit Analysis in a New Zealand public sector context, with particular emphasis on the basic questions likely to arise. (G,N)	NZ Dept of Treasury
	What are the Tools for Economic Appraisal	Brisbane City Council (G,N)	Brisbane, Australia
	SCRAPS - Sewer cataloguing, retrieval, and prioritization system	This software tool helps users identify pipelines at risk of structural and operational failure. (T,C)	WERF

Asset Management Element	Tool Name	Description[a]	Source
	SEESAW	financial and economic modeling software	WRc
	TR 61	Capex, opex models for estimating costs (T,C)	WRc
	SIMPLE – Sustainable Infrastructure Management Program Learning Environment.	Web-based knowledge base provides state-of-the-art training, guidance, research, and tools for AM practitioners at utilities of all types, sizes, and levels of AM. (T,C)	WERF
	Prime-Choice Decision Support Method (trademarked)	combines financial and risk assessment to aid in selection among a set of competing alternatives (T,C)	LMI Govt Consulting
	International Infrastructure Management Manual	Starts with basic asset management principles to practical steps for implementing advanced asset management systems (G,C)	INGENIUM
Funding Requirements	Capital Asset Plan and Business Case Summary	(T,N)	U.S. EPA
	CAP Finance	Icon-driven capital asset inventory and financial planning software program (T,N)	Boise State EFC
	TEAMS – Total Electronic Asset Management System	AM software for small to medium wastewater utilities (T,N)	U.S. EPA
	Building Life Cycle Costing (BLCC) Program	Software to provide computational support for the analysis of capital investments in buildings (T,N)	US DOE
	Sources of Technical & Financial Assistance for Small Drinking Water Systems	Listing and description of financing organizations and technical assistance providers (G,N)	U.S. EPA
	SIMPLE – Sustainable Infrastructure Management Program Learning Environment.	Web-based knowledge base provides state-of-the-art training, guidance, research, and tools for AM practitioners at utilities of all types, sizes, and levels of AM. (T,C)	WERF
	International Infrastructure Management Manual	Starts with basic asset management principles to practical steps for implementing advanced asset management systems (G,C)	INGENIUM
Major Failure Modes	Standardized System to Assess the Condition of Infrastructure in Edmonton	Based on five-point scale across three failure criteria – level of service, physical mortality, and capacity (T,N)	City of Edmonton, CA
	ProRIsk	Examines failure modes and differing investment needs (T,C)	WRc
	SIMPLE – Sustainable Infrastructure Management Program Learning Environment.	Web-based knowledge base provides state-of-the-art training, guidance, research, and tools for AM practitioners at utilities of all types, sizes, and levels of AM. (T,C)	WERF
Asset Management Planning	Strategic Planning: A Handbook for Small Water Systems	Basic planning with worksheets (non-electronic) (G,C)	U.S.EPA
	Continual improvement in utility management: A framework for integration	Roadmap showing how the management initiatives interrelate and how a utility can best approach integrating them in the context of a continual improvement management system framework. (G,C)	NACWA
	International Infrastructure Management Manual	Starts with basic asset management principles to practical steps for implementing advanced asset management systems (G,C)	INGENIUM
Training	TEAMS – Total Electronic Asset Management System	AM software for small to medium wastewater utilities (T,N)	U.S. EPA

Asset Management Element	Tool Name	Description[a]	Source
	U.S. EPA Asset Management Workshop Training Material	Advanced Asset Management two day training workshop presentations (G,N)	U.S.EPA
	SIMPLE – Sustainable Infrastructure Management Program Learning Environment.	Web-based knowledge base provides state-of-the-art training, guidance, research, and tools for AM practitioners at utilities of all types, sizes, and levels of AM. (T,C)	WERF
	International Infrastructure Management Manual	Starts with basic asset management principles to practical steps for implementing advanced asset management systems (G,C)	INGENIUM

[a]Tool (T), Guide (G), Non-Commercial (N), Commercial (C)

2.5 Conclusion

This study has identified a limited number of non-commercial tools to aid in asset management decision-making. Many of the tools that were identified have been developed for types of infrastructure other than water services, such as roads or facilities. Despite that, a few of the tools may have limited applicability to the water industry, but could not be considered as being customized for water users.

The study has also identified a range of commercially available tools – some of which are directly related to water utilities, as well as a number of guidelines that are available on both a non-commercial and a commercial basis. References to tools developed for other forms of infrastructure have been included as it may be less costly for them to be redesigned for the water industry than to be developed from first principles. A separate study will be undertaken to determine the usefulness of the identified asset management tools to WERF and Water Research Foundation members.

CHAPTER 3.0

DECISION SUPPORT TOOLS: PRIORITIES FOR DEVELOPMENT

The WERF Strategic Asset Management research program includes a goal for Decision Support Tools and Implementation Guidance, organized around two major elements:

Analytic/Decision Support Tools – Develop staged, user-friendly analytic and decision support tools (gap analysis, cost benefit analysis, risk management, life cycle costing, condition assessment selection tool, etc.) for SAM implementation and continuous improvement.

Implementation Guidance– Provide practical, accessible guidance on how to implement SAM in incremental stages leading to a well-planned, structured, and progressive implementation to suit the needs of a diverse range of utilities' communication practices.

An important deliverable for the project is a prioritized set of SAM decision tools recommended for development. These tools would be organized and developed within the core asset management conceptual framework embedded in SIMPLE, WERF's on-line knowledge base for asset management (Sustainable Infrastructure Management Program Learning Environment). The tools, instructions for their use and deployment, and case studies describing their application would all be added to the SIMPLE knowledge base.

3.1 Methodology

GHD developed a draft set of core SAM tools with general descriptions of each. These tools directly link to and provide support for the "Ten Step Process" and the "Five Core Questions" that form the heart of the SIMPLE conceptual framework. The draft list of identified tools was presented to a working group comprised of practitioners from participating utilities in January 2008. (The workshop agenda is attached as Appendix B.) This group was asked, through facilitation, to:

- review and discuss the nature of each tool;
- recommend additional tools for the list;
- remove any that seemed of little value to utility practitioners;
- indicate to which of three levels the tools should be initially targeted (with Level 1 being very basic and Level 3 being advanced); and
- prioritize the tools by level such that the creation of the tools could be grouped into phases.

Each tool on the resultant list was then "scoped" at a high level by GHD; that is, each tool was described, its level and priority reflected, the benefits to a user identified, the deliverables enumerated, the tasks described and an estimate of hours and costs developed. This document presents that prioritization and resultant scoping as shown in Table 3-1.

Table 3-1. Minimum Set of Core Asset Management Tools.

Recommended Tool	Features
What is the state of my assets?	
Default Asset Hierarchy Tool (water and wastewater)	Provides a working hierarchical structure for a large, full-service water and wastewater utility reaching down five levels
Condition Assessment Tool	Use WERF's Condition Assessment Protocol results (MWH); modify for water assets, add Levels 1 & 2
Residual Life Tool	Systematically converts condition scores to an estimate of maximum & effective physical residual life and economic residual life, based on various levels of data
	Develop on-line baseline table of design lives and supporting database contributed by participating agencies to over time validate Mean Time Between Failure (MTBF)
Life Cycle Costing Tool	Systematically projects life cycle costs under various failure modes and projected costs trends by major cost category
Valuation Tool	Structures the systematic valuation of assets based on several increasingly sophisticated techniques
Failure Modes Tool (checklist)	Guides in the determination of Most Imminent Major Failure Mode and associated relevant management strategy
	Facilitates determination of tactical level failure modes for Physical Mortality Major Mode
What sustained Level of Service/performance is required?	
LOS Tool (checklist)	A checklist for creating a Triple Bottom Line oriented, organization level Statement of Levels of Service
	Demonstrates how to set asset performance levels that link to the organization's strategic LOS statements
Which assets are critical to sustained performance?	
Risk/consequence (Business Risk Exposure) Analysis / Risk management tools	Presents increasingly sophisticated methods to set Business Risk Exposure Scores (probability of failure versus consequence of failure matrix), ranging from simple scores to full economic costs.
	Facilitates determining and weighting strategic risk to the organization across all aspects of the business
Confidence Level Ranking Tool	Assigns scores to management "treatment" alternatives based on a weighted "value chain" of elements that determine the robustness of the evaluated treatment solution

Table 3-1. Minimum Set of Core Asset Management Tools, continued.

Tools	Features
Given the above, what are my best maintenance & capital investment options?	
Benefit/Cost analysis / "Triple Bottom Line" Business Case Tool	See separate proposal for Benefit/Cost tool.
	Business Case Tool assists in evaluating from a business metric perspective the relative investment merits of alternative treatment solutions, whether O&M or capital in nature based on Triple Bottom Line
Optimized Renewal Strategies Tool	Assists in determining which renewal strategy (maintenance/repair, refurbishment, replacement) to pursue and when to transition based on the life cycle of a specific asset, a class of assets, or the entire portfolio of assets
Maintenance Strategies / RCM Tool	Assists in the systematic determination of which set of maintenance tactics (reactive, proactive) are most cost effective for a specific asset based on reliability centered maintenance concepts
CIP Validation Tool	Assists in determining when a proposed capital solution is sufficiently sound to move to investment funding stage
	Assigns a confidence level rating to each project/solution that identifies what aspects of a proposed project/solution need strengthening to increase confidence in the proposed solution
CIP Prioritization Tool	Once the project list is validated, this tool assists in prioritization of projects so that a specific CIP list can be adopted

What funding strategy is required?	
Funding Elements Tool (checklist and examples)	Relates "optimized" investment decisions made through other tools to projected budget, rate, and financial condition impacts
	Advanced stages could allow scenario modeling that suggests optimized investments are given as a defined LOS, condition/residual life data, business risk exposure scores and various "available budget" target levels

3.2 Prioritization by Levels of the SAM Tools

The working group voted on levels of development and prioritization for each tool in the workshop held January 30-31, 2008 at WERF headquarters in Alexandria, VA. Table 3-2 summarizes the results of the balloting.

Table 3-2. Prioritization Levels of SAM Tools.

	Priority (Phase)			Initial Level		
Question	1	2	3	1	2	3
What is the state of my assets?						
Asset Hierarchy Tool	11	3	0	7	2	0
Condition Assessment Tool	11	3	0	1	8	3
Residual Life Tool	7	4	2	5	9	2
Life Cycle Costing Tool	8	3	1	2	11	0
Valuation Tool	5	3	4	9	3	0
Failure Modes Tool (checklist)	9	1	2	5	6	2
What sustained Level of Service/performance is required?						
LOS Tool (checklist)	8	4	1	5	6	0
Which assets are critical to sustained performance?						
Risk/consequence (Business Risk Exposure) Analysis / Risk management tools	7	5	2	2	11	1
Confidence Level Ranking Tool	4	5	3	5	6	0
Given the above, what are my best maintenance & capital investment options?						
Benefit/Cost analysis / "Triple Bottom Line" Business Case Tool	5	6	2	2	9	1
Optimized Renewal Strategies Tool	4	7	3	4	8	1
Maintenance Strategies / RCM Tool	2	4	5	7	4	1
CIP Validation Tool	6	6	1	5	7	2
CIP Prioritization Tool	5	6	2	5	7	2
What funding strategy is required?						
Funding Elements Tool	2	3	7	4	5	1
Nessie Tool (Optimized)	0	3	8	4	2	4
Nessie Tool (Not Optimized)	3	6	3	3	7	1
Asset Summary Tool	3	2	2	2	6	1

Based on indications of priority by the working group in workshop setting and on interconnectivities among the various Tools (certain tools build on input from other tools), the following prioritization structure is recommended as a guide to tool development (Table 3-3).

Table 3-3. Prioritization Structure.

Question	Phase 1			Phase 2			Phase 3		
	Level 1	Level 2	Level 3	Level 1	Level 2	Level 3	Level 1	Level 2	Level 3
What is the state of my assets?									
Asset Hierarchy Tool	Phase 1								
Condition Assessment Tool	Phase 1	Phase 2							
Residual Life Tool	Phase 1	Phase 2							
Life Cycle Costing Tool	Phase 1	Phase 2							
Valuation Tool	Phase 1	Phase 2							
Failure Modes Tool (checklist)	Phase 1	Phase 2							
What sustained Level of Service/performance is required?									
LOS Tool (checklist)	Phase 1	Phase 2							
Which assets are critical to sustained performance?									
Risk/consequence (Business Risk Exposure) Analysis / Risk management tools	Phase 1	Phase 2							
Confidence Level Ranking Tool				Phase 1	Phase 2				
Given the above, what are my best maintenance & capital investment options?									
Benefit/Cost analysis / "Triple Bottom Line" Business Case Tool	Phase 1	Phase 2							
Optimized Renewal Strategies Tool				Phase 1	Phase 2				
Maintenance Strategies / RCM Tool							Phase 1		
CIP Validation Tool	Phase 1	Phase 2							
CIP Prioritization Tool				Phase 1	Phase 2				
What funding strategy is required?									
Funding Elements Tool							Phase 1	Phase 2	
Nessie Tool (Optimized)							Phase 1	Phase 2	Phase 3
Nessie Tool (Not Optimized)				Phase 1	Phase 2				
Asset Summary Tool	Phase 1								

KEY:
Phase 1 (hatched)
Phase 2 (grey)
Phase 3 (black)

3.3 Scopes of the Recommended SAM Tools

The input received from the general discussion and the results derived from the systematic voting from the January 2008 workshop have been used to develop a more detailed scope for each recommended tool. The scope for each tool treats the development of each as a stand alone development project. This allows WERF to "mix and match" tool development according to available funding and funding sources while keeping the general phase and level largely intact.

Table 3-4 below summarizes the proposed scope and projected development cost (excluding project management), including expenses for each tool. Additional descriptive material related to each tool follows the table.

Table 3-4. Tool Summary.

Question/ Tool Name	Description	Phase	Level	Projected Cost ($)
What is the state of my assets?				
AR1 - Asset Hierarchy/Registry Tool	Five level asset hierarchy for mid-sized wastewater system in Excel or Tree Pad format	1	1	32,700
AR2- Asset Hierarchy/Registry Tool	Twelve level asset hierarchy for mid-sized wastewater system in Excel or Tree Pad format	1	2	30,200
CA1 – Condition Assessment Tool	Single attribute condition classification tool in Excel table format	1	1	38,220
CA 2 -Condition Assessment Tool	Multi –attribute Excel condition rating tool for up to four classes of assets	1	2	101,415
RL1 – Residual Life Tool	Procedure and Excel table to determine remaining useful physical life based on age /design life	1	1	36,150
RL2 - Residual Life Tool	Procedure and Excel table to determine useful life based on level 1 modified by local factors	1	2	31,320
LCC2 - Life Cycle Costing Tool	Excel table enabling organization and collection of life cycle cost data and projection of cost trends for assets	1	1 & 2	47,880
RC1 – Replacement Cost Valuation Tool	Excel workbook estimating replacement costs using historical costs, indices, and location factors	1	1	47,880
RC2 – Replacement Cost Valuation Tool	Excel workbook estimating replacement costs based on std unit costs, with inflation indexing	1	2	114,120
FM - Failure Modes Tool (checklist)	Procedural checklist to assist in identifying the imminent failure mode for an asset.	1	1	15,450
What sustained Level of Service/performance is required?				
LOS 1 –Level of Service Tool (checklist)	A checklist to assist in setting levels of service on a triple bottom line basis, and associated performance measures	1	1	15,450
Which assets are critical to sustained performance?				
BRE 1 - Risk/consequence (Business Risk Exposure) Analysis / Risk management tools	Excel worksheet to determine a BRE score based on a single aspect scale for probability and consequences of failure.	1	1	31,320
BRE 2 - Risk/consequence (Business Risk Exposure) Analysis / Risk management tools	Excel worksheet to determine a level 2 BRE score using multi-aspect scales for probability and consequences of failure, and the level of redundancy	1	2	47,880
BRE 3 - Risk/consequence (Business Risk Exposure) Analysis / Risk management tools	Excel worksheet to determine a level 3 BRE score based on the "Americanization" of an existing UKWIR Risk tool	1	3	58,230
CLR 2 - Confidence Level Ranking Tool	Excel spreadsheet that facilitates a level 1 rating of the level of confidence in decision-making	2	2	47,880

Table 3-4. Tool Summary, continued.

Question/ Tool Name	Description	Phase	Level	Projected Cost ($)
Given the above, what are my best maintenance & capital investment options?				
BC 2 - Benefit/Cost analysis / "Triple Bottom Line" Business Case Tool	Excel workbook that assists in the analysis of feasible investment options in support of a business case for a decision.	1	2	47,880
ORDM 3 - Optimized Renewal Strategies Tool	An Access application that facilitates the selection of the most cost effective asset renewal strategy.	2	3	114,120
RCM 2 - Maintenance Strategies / RCM Tool	An Access application that facilitates selection of the most cost effective maintenance strategy for an asset	3	2	114,120
CIP V2 - CIP Validation Tool	An Excel workbook that facilitates the analysis of feasible investment options for Capital Improvement Projects.	1	2	47,880
CIP 2 - CIP Prioritization Tool	An Excel workbook designed to determine CIP prioritization scores based on a weighting of selected criteria.	2	2	27,180
What funding strategy is required?				
FE 2 - Funding Elements Tool	Excel workbook enabling assessment of the impacts of various levels of investment on future budget, rate and financial conditions.	3	2	58,230
Renewal Cost Projection –"Nessie Curve" Tool (Optimized)	An Excel workbook that graphically depicts projected renewal costs over a period of time.	3	1	101,415
Nessie Tool (Not Optimized)				
Asset Summary Tool				

The scope lists the recommended phase and level of the described tool. Each tool is described and the benefits that should be expected are reviewed. The deliverables for the tool project are articulated with the general tasks necessary to develop the tool being listed. Where applicable, interconnectivity to other tools is noted (which has implications for phasing), and an estimated budget is presented for development of each of the tools.

A general scope for each of the recommended SAM tools is presented in this section.

3.4 Tool: AR1 – Asset Hierarchy/Registry

Priority: Phase 1 – Level 1

Description

This tool project provides two working examples of a hierarchical asset register structure reaching down to a maximum of five levels, one each for a mid-sized, full-service water utility and a mid-sized, full service wastewater utility. . The tool will serve as a starting point for a utility to develop its own hierarchy

A beta prototype register will be developed in Excel and Tree Pad for water and a wastewater utility. A Practitioner Review Team will review beta prototypes modify as relevant. Two example case studies of up to six pages each – one each water and wastewater will be prepared in addition to developing user instructions. The final versions will be incorporated into SIMPLE.

Benefits

The asset register is the foundation of asset management. Best practice calls for the register to be organized in a hierarchical structure such that child assets can be rolled up to parent levels, then to grandparent level, etc. This roll-up capability allows the practitioner to focus analysis and decision-making at any relevant level in the organization from the individual asset, to the major process, to the facility, the system, and the network and eventually the enterprise level.

Interconnectivities/Interdependencies

No direct interdependencies with other tools.

3.5 Tool: AR 2 – Asset Hierarchy/ Registry

Priority: Phase 1 – Level 2

Description

This tool project provides two working examples of a hierarchical asset register structure reaching down to a maximum of 12 levels, one each for a large, full-service water utility and a large, full-service wastewater utility.

A beta prototype register will be developed in Excel and Tree Pad for water and a wastewater utility. A Practitioner Review Team will review beta prototypes modify as relevant. Two example case studies of up to six pages each – one each water and wastewater will be prepared in addition to developing user instructions. The final versions will be incorporated into SIMPLE.

Benefits

The asset register is the foundation of asset management. Best practice calls for the register to be organized in a hierarchy such that child assets can be rolled up to parent levels, then to grandparent, etc. This roll-up capability allows the practitioner to focus analysis and decision making at any relevant level in the organization from the individual asset, to the major process, to the facility, the system, and the network and eventually the enterprise level.

Interconnectivities/Interdependencies

Directly linked to AR 1 and leverages those associated Task Elements.

3.6 Tool: CA 1 – Condition Assessment

Priority: Phase 1 – Level 1

Description

This tool provides for a basic level characterization of condition for all types of assets through use of a generic, single aspect coding structure. The tool will be a simple single-attribute, generic scale for classifying the condition of virtually any class of water/wastewater assets. Additionally, a prototype condition assessment protocol will be incorporated outlining aspects of mobilization for and execution of a basic level condition assessment program

A beta prototype Excel table for classifying and recording the condition of assets and a condition assessment protocol will be developed. A Practitioner Review Team will review beta prototypes modify as relevant. Two example case studies of up to six pages each – one each water and wastewater will be prepared in addition to developing user instructions. The final versions will be incorporated into SIMPLE.

Benefits

The condition of an asset provides valuable insight into its remaining useful physical life. It is an essential piece of information that ultimately assists in identifying what types of management strategies to deploy and when to transition from one strategy to another. Best Practice dictates that all assets should have at least a basic level of condition assessment on a regular, periodic basis. The assessment should be systematic and consistently applied.

Interconnectivities/Interdependencies

No direct interconnectivities or interdependencies.

3.7 Tool: CA 2 – Condition Assessment

Priority: Phase 1 – Level 2

Description

These tools provide for an intermediate level characterization of condition for all types of assets through use of various multi-dimensional coding structures targeted to major classes of assets.

This phase will produce a multi attribute condition rating tool in Excel that features up to four major classes of assets (for example, buried passive assets such as pipes, buried active assets such as valves, above ground passive assets such as tanks and pipes, and above ground active assets such as motors and pumps), where the major attributes for each are relevant to that major class (for example, corrosion might be a rated attribute for the underground passive class); incorporate metrics that can reflect current performance and reliability in addition to condition. The effort will include a prototype condition assessment protocol outlining aspects of mobilization for and execution of a condition assessment program for each major class

A Practitioner Review Team will review beta prototypes and modify as relevant. Two example case studies of up to six pages each – one each water and wastewater will be prepared in addition to developing user instructions. The final versions will be incorporated into SIMPLE.

Benefits

The condition of an asset provides valuable insight into its remaining useful physical life. It is an essential piece of information that ultimately assists in identifying what types of management strategies to deploy and when to transition from one strategy to another. While a basic level of condition rating is essential to good asset management, more advanced levels of condition assessment lead to higher confidence in decisions driven by condition of assets.

Interconnectivities/Interdependencies

Directly linked to CA 1 and leverages those associated task elements.

3.8 Tool RL1 – Residual Life

Priority: Phase 1 – Level 1

Description

This tool provides for the systematic estimation of residual physical life using age and design life of the asset as the drivers.

A systematic procedure and Excel structure to guide the determination of remaining useful physical life based on age of the asset and expected design life will be developed. The tool will include a table of "default" useful lives for a wide range of assets that could be used in the absence of better local data.

A Practitioner Review Team will review beta prototypes and modify as relevant. Two example case studies of up to six pages each – one each water and wastewater will be prepared in addition to developing user instructions. The final versions will be incorporated into SIMPLE.

Benefits

One of the most important pieces of information about an asset for purposes of determining the lowest life cycle cost investment in that asset is a determination of its remaining useful life. The most basic technique for determining remaining life uses the age of the asset and an estimate of its expected design life. By assigning a systematic estimate of remaining useful lives across the portfolio, an agency can build a reasonable projection of future renewal funding requirements over a defined period of time.

Interconnectivities/Interdependencies:

None.

3.9 Tool RL 2 – Residual Life

Priority: Phase 1 – Level 2

Description

This tool provides for the systematic estimation of residual physical life using age and design life of the asset as modified by local factors as the drivers. The tool would incorporate the table of "default" useful lives developed in RL 1.

A Practitioner Review Team will review beta prototypes and modify as relevant. Two example case studies of up to six pages each – one each water and wastewater will be prepared in addition to developing user instructions. The final versions will be incorporated into SIMPLE.

Benefits

As noted in RL 1, one of the most important pieces of information about an asset for purposes of determining the lowest life cycle cost investment in that asset is a determination of its remaining useful physical life. The most basic technique for determining remaining life uses the age of the asset and an estimate of its expected design life. Level 2 provides for a systematic modification of the design life by introducing local factors (e.g. operating environment, quality/level of design, etc) that can be expected to increase or decrease asset lives. By assigning a systematic estimate of remaining useful lives across the portfolio, an agency can build a reasonable projection of funding requirements over a defined period of time.

Interconnectivities/interdependencies:

Directly linked to RL 1 and leverages those associated task elements.

3.10 Tool LCC2 – Life Cycle Costing

Priority: Phase 1 – Levels 1 and 2

Description

This tool facilitates the systematic organization and collection of historic life cycle costs by major cost category for an asset or class of similar assets, assists in identifying cost trends, and in projecting future costs. (Since level 1 in this case would simply be a reduced version of level 2, this tool is more cost effectively developed as a single tool at level 2.)

This tool will consist of an Excel table that 1) facilitates the organization and collection of life cycle costs for an asset or group of assets and 2) provides for the systematic identification of cost trends and the projection of those costs over a defined time frame.

The researcher will develop a beta prototype Excel based structure for organizing and collecting life cycle costs for a wide range of water and wastewater utility type assets and a prototype life cycle cost protocol. A Practitioner Review Team will review beta prototypes and modify as relevant. Two example case studies of up to six pages each – one each water and wastewater will be prepared in addition to developing user instructions. The final versions will be incorporated into SIMPLE.

Benefits

Understanding life cycle costs is a key component of effectively managing assets to achieve lowest total cost of ownership investment. Life cycle costs and renewal valuation ("replacement costs") combine to establish the financial framework within which investment decisions (maintenance as well as capital) are made.

Interconnectivities/Interdependencies

None.

3.11 Tool RC 1 – Replacement Cost

Priority: Phase 1 – Level 1

Description

This tool facilitates the estimation of replacement costs based on historic costs of the same or equivalent assets.

This Excel workbook will facilitate the estimation of replacement costs for a wide range of utility assets based on historic cost and date of acquisition; in addition to general inflation indexing, relevant construction related cost indexes will be provided for estimation purposes. Greenfield to Brownfield conversion cost factors will be incorporated.

The researchers will develop a beta prototype Excel based structure for estimating replacement costs based on historic cost and date of acquisition and a prototype replacement cost protocol. A Practitioner Review Team will review beta prototypes and modify as relevant. Two example case studies of up to six pages each – one each water and wastewater will be prepared in addition to developing user instructions. The final versions will be incorporated into SIMPLE. Benefits

As noted in the Life Cycle Costing tools, replacement cost is a key element of investment decision-making. (Replacement costs are one element of renewal costs; the other two elements include repair costs and refurbishment costs. In a fully robust approach to decision-making, costs for all three elements should be determined so that optimized investment analysis can be executed and business cases developed. Because not all assets can be repaired or refurbished, this level of tool focuses only on replacement cost estimation).

Interconnectivities/Interdependencies

None.

3.12 Tool RC2 – Replacement Costs

Priority: Phase 1 – Level 2

Description

This tool facilitates the estimation of replacement costs where actual historic costs of the asset to be replaced are not known.

The research will develop an Excel workbook that facilitates the estimation of replacement costs, based on historic cost and date of acquisition, for a wide range of utility assets based on standard unit costs for major process or sub system level assets using the asset hierarchy developed in AR 1; in addition to general inflation indexing, relevant construction related cost indexes will be provided for estimation purposes. Greenfield to Brownfield conversion cost factors will be incorporated.

A Practitioner Review Team will review beta prototypes and modify as relevant. Two example case studies of up to six pages each – one each water and wastewater will be prepared in addition to developing user instructions. The final versions will be incorporated into SIMPLE.

Benefits

As noted in the Life Cycle Costing tools, replacement cost is a key element of investment decision making. (Replacement costs are one element of renewal costs; the other two elements include repair costs and refurbishment costs. In a fully robust approach to decision-making, costs for all three elements should be determined so that optimized investment analysis can be executed and business cases developed. Because not all assets can be repaired or refurbished, this tool focuses first on replacement cost estimation).

Interconnectivities/Interdependencies

Directly linked to RC 1 and leverages those associated Task Elements.

3.13 Tool FM Primary Failure Modes (Checklist)

Priority: Phase 1 – Level 1

Description

All assets will eventually fail. Failure occurs when an asset cannot do what the user requires it to do in its operating environment. Most think of failure as being driven by physical "collapse". In fact, physical collapse or "mortality" is but one major mode of failure. Three other high level or primary failure modes are recognized – each with significantly different implications for cost effective management. The four primary failure modes are: 1) capacity, 2) level of service, 3) physical mortality, and 4) efficiency. All four major failure modes operate on all assets at all times. The significant questions in front of management are – which of the four is most imminent and when is it likely to fail under that primary failure mode?

This tool will be a procedural checklist that facilitates the identification of the imminent primary failure mode. A Practitioner Review Team will review beta prototypes and modify as relevant. Two example case studies of up to six pages each – one each water and wastewater will be prepared in addition to developing user instructions. The final versions will be incorporated into SIMPLE.

Benefits

Because each failure mode has such substantially different implications for cost effective management of the asset, it is important to identify the most imminent failure mode likely to impact a given asset. This allows the practitioner to develop appropriate management strategies to fit the most imminent mode and to time the interventions accordingly.

Interconnectivities/Interdependencies

Ties to residual life tools (RL 1 and RL 2).

3.14 Tool LOS 1 – Level of Service (Checklist)

Priority: Phase 1 –Levels 1 and 2

Description

This tool is a procedural checklist that assists a utility to establish (and periodically measure) targeted and actual levels of service, both at the enterprise (strategic) level and at the asset (operations) level. The tool will be developed using a "triple bottom line" approach (that is, a financial, environmental, and community/organizational perspective). Because level 1 is a direct simplification of level 2, only a level 2 tool is required.

A Practitioner Review Team will review beta prototypes and modify as relevant. Two example case studies of up to six pages each – one each water and wastewater will be prepared in addition to developing user instructions. The final versions will be incorporated into SIMPLE.

Benefits

All of the "best practices" associated with asset management can be characterized as directed at assisting the utility management team to sustain performance at the lowest total cost of ownership. Consequently, before a utility can determine whether it is being successful in achieving the lowest total (life cycle) cost of sustaining performance, just what "performance" means must be defined. Once performance at the strategic level has been targeted, what that strategic level target means in terms of asset performance at the plant, major process, sub-system and ultimately the asset level must be understood if the strategic vision is to be actually achieved.

Interconnectivities/Interdependencies

No direct links.

3.15 Tool BRE 1 Business Risk Exposure

Priority: Phase 1 – Level 1

Description

This tool is the basic level of three levels of tools that assist the practitioner in characterizing the business risk exposure (BRE) of the utility associated with assets through assigning risk scores to assets or groups of assets. As a level 1 tool, it incorporates a generic, single-aspect scale for characterizing the probability of failure and a generic, single aspect scale for the consequences of failure. A third element, redundancy, is included as a simple multiplicative factor to the core expression of probability time's consequence.

The tool will be a beta prototype Excel based structure for determining a basic Business Risk exposure score for a single asset or group of closely related assets. A prototype replacement cost protocol will be developed as part of this research.

A Practitioner Review Team will review beta prototypes and modify as relevant. Two example case studies of up to six pages each – one each water and wastewater will be prepared in addition to developing user instructions. The final versions will be incorporated into SIMPLE.

Benefits

Business risk exposure is a metric that relates the probability of failure of an asset or group of assets with the consequence of failure to the utility. The Business Risk Exposure score is a major input in determining toward which assets to focus maintenance and capital investment. Best practice informs that all assets should have at least a level 1 BRE score.

Interconnectivities/interdependencies:

Uses insights gleaned from condition assessment.

3.16 Tool BRE 2 – Business Risk Exposure

Priority: Phase 1 – Level 2

Description

This tool is level 2 of three levels of tools that assist the practitioner in characterizing the business risk exposure of the utility associated with assets through assigning risk scores to assets or groups of assets. As a level 2 tool, it incorporates a multi-aspect scale for characterizing the probability of failure and a multi-aspect scale for the consequences of failure. The third element, redundancy, is expanded to incorporate factors for a range of mitigation strategies.

A Practitioner Review Team will review beta prototypes and modify as relevant. Two example case studies of up to six pages each – one each water and wastewater will be prepared in addition to developing user instructions. The final versions will be incorporated into SIMPLE.

Benefits

Level 2 BRE scores provide more depth for understanding the potential business risk exposure that a utility faces due to the failure of a specific asset or group of closely related assets. Assets with low level 1 scores do not need level 2 analysis.

Interconnectivities/Interdependencies

Directly linked to BRE Tools and leverages those associated Task Elements.

3.17 Tool BRE 3 – Business Risk Exposure (BRE)

Priority: Phase 1 – Level 3

Description

This tool is level 3 of three levels of tools that assist the practitioner in characterizing the business risk exposure of the utility associated with assets through assigning risk scores to assets or groups of assets. As a level 3 tool, it incorporates more detail in the generation of probability of failure and consequence of failure scores. Risk is reported in terms of probability weighted economic value (probability multiplied by the estimated economic consequences of failure). This tool has been developed by UKWIR for application in the UK. WERF has negotiated rights to the tool. Most of the effort will be invested in "Americanizing" the tool and in incorporating it into SIMPLE.

The research team will develop a beta prototype Excel based structure for determining a level 3 Business Risk exposure score for a single asset or group of closely related assets by adapting the UKWIR Risk Tool to the North American water industry, as described above. A prototype methodological protocol will be developed as part of the tool.

A Practitioner Review Team will review beta prototypes and modify as relevant. Two example case studies of up to six pages each – one each water and wastewater will be prepared in addition to developing user instructions. The final versions will be incorporated into SIMPLE.

Benefits

Level 3 BRE scores provide even more depth for understanding the potential business risk exposure that a utility faces due to the failure of a specific asset or group of closely related assets. Because of its sophistication and level of effort, it should be used for investment decisions involving high capital investment.

Interconnectivities/Interdependencies

Directly linked to BRE 1 and BRE 2 and leverages those associated Task Elements.

3.18 Tool CLR2 – Confidence Level Ratings

Priority: Phase 2 – Level 2

Description

This tool assists the practitioner in assigning scores to management "treatment" alternatives based on a weighted "value chain" of elements that determine the robustness of the evaluated treatment solution. The score is a metric from 0 to 100. The resultant score is an indication of the depth of analysis and the quality of data and information that has been used in generating a recommended investment.

The tool will be an Excel worksheet that facilitates the determination of a level 1 Confidence Level Rating score for a single asset or group of closely related assets, as outlined above. A Practitioner Review Team will review beta prototypes and modify as relevant. Two example case studies of up to six pages each – one each water and wastewater will be prepared in addition to developing user instructions. The final versions will be incorporated into SIMPLE.

Benefits

A utility can be said to confront two distinctly different types of business risk – risk associated with the probability and consequences of failure of its assets and risk associated with deploying poor investment decisions (including investment too early or too late in the life cycle). The scale of the latter risk is proportionate to the level of investment involved. Confidence level ratings offer a systematic benchmark against which decision makers can weigh the cost of obtaining better information against the urgency of a given situation. It is a powerful tool for both operations and capital budgeting.

Interconnectivities/Interdependencies

No direct links.

3.19 Tool BC 2 – Business Case Analysis

Priority: Phase 1 – Level 2

Description

This tool assists the practitioner in evaluating, from a business metric perspective, the relative investment merits of a range of alternative management solutions, whether operations, maintenance, or capital. To add further insight, the tool incorporates an optional "Triple Bottom Line" perspective – that is, it facilitates analysis from not only a financial perspective, but also an environmental and community perspective, if desired. Alternative management solutions would include such options as do nothing, status quo, change operations, change maintenance, repair, refurbish, replace, and decommission/reassign as feasible.

The tool will be a beta prototype Excel based structure for the analysis of feasible investment options for an asset or group of assets and that assists in preparing a concise business case for the proposed decision as described above. A Practitioner Review Team will review beta prototypes and modify as relevant. Two example case studies of up to six pages each – one each water and wastewater will be prepared in addition to developing user instructions. The final versions will be incorporated into SIMPLE.

Benefits

Staff routinely recommends to its relevant policy board the investment of millions of dollars in its infrastructure. This investment ranges in form from operations to maintenance to capital. In an era of increasing rate resistance, it is incumbent upon staff to establish that the recommendation submitted for consideration is indeed the very best investment option available given the decision context. This tool facilitates the systematic development of a business case that substantiates each investment recommendation. Such substantiation increases confidence in the decision on the part of both staff and board (thereby breaking down tendencies toward adversarial relationships between staff and board) and tends to strengthen credibility with all stakeholders.

Interconnectivities/Interdependencies

Extensive interconnectivity with Business Risk Exposure, Confidence Level Rating, and asset replacement costing tools.

3.20 Tool ORDM 3 – Optimized Renewal Decision Making Strategies

Priority: Phase 2 – Level 3

Description

This tool assists the asset manager in determining which renewal strategy (maintenance/repair, refurbishment, replacement) is most cost effective and when in the life cycle of a given asset to transition from an operations and maintenance strategy to a capital investment (renewal) strategy given a set of user established baseline conditions and assumptions.

Researchers will develop a beta prototype Access based application for the analysis of feasible investment options for an asset or group of assets and that assists in preparing a concise business case for the proposed decision as described above. A Practitioner Review Team will review beta prototypes and modify as relevant. Two example case studies of up to six pages each – one each water and wastewater will be prepared in addition to developing user instructions. The final versions will be incorporated into SIMPLE.

Benefits

All assets deteriorate with usage and over time – in effect, all assets are progressing toward failure. Asset managers can choose from three different forms of investment to sustain performance from a deteriorating asset – operations, maintenance, or capital. In reality, each asset requires a mix of all three forms of investment to sustain long term performance. There exists, then, a theoretical optimum mix of operations, maintenance, and capital investment over the life of the asset that leads to the lowest total cost of ownership for a given level of service. Optimized renewal decision making is a set of user established "business rules" which guide the proportionate balance and timing of the three types of intervention investment in the search for the lowest total cost of ownership. The business rules that drive optimization include interactions among age, condition, consequence of failure, residual life, operations cost trends, and maintenance cost trends, deterioration curves, and renewal costs.

Interconnectivities/Interdependencies

This tool builds on all of the other tools in the recommended toolkit.

3.21 Tool RCM 2 – Maintenance Strategies/RCM

Priority: Phase 3 – Level 2

Description

This tool assists the maintenance manager in the systematic determination of which set of maintenance tactics (reactive, proactive) are most cost effective for a specific asset based on reliability centered maintenance concepts.

The tool will be a beta prototype Access-based application for the determining the most cost effective maintenance strategy for a given asset or group of assets as described above. A Practitioner Review Team will review beta prototypes and modify as relevant. Two example case studies of up to six pages each – one each water and wastewater will be prepared in addition to developing user instructions. The final versions will be incorporated into SIMPLE.

Benefits

Maintenance is a major form of investment on the part of virtually every utility. But not all maintenance is cost effective. Private sector research indicates that as much as 70% of the typical maintenance regimen is not cost effective. Certain assets are most cost-effectively run to failure; for others, that is the worst strategy. Preventive maintenance can be overdone, especially if it is not integrated with predictive techniques to ascertain the most cost effective timing of preventive strategies. Reliability Centered Maintenance is a discipline built around a set of analytics that assists the practitioner to determine which of many options is the most cost-effective maintenance strategy.

Interconnectivities/Interdependencies

Condition assessment, Business Risk Exposure, replacement costing tools.

3.22 Tool CIP V2 – Validation

Priority: Phase 1 – Level 2

Description

This tool assists the asset manager in determining when a proposed capital investment is sufficiently sound to move to a budget funding stage by facilitating the systematic assignment of a confidence level rating to each project/solution. The confidence level rating is a metric that not only rates the soundness of the capital investment but it also identifies what aspects of a proposed project/solution need strengthening to increase confidence in the proposed solution. The methodology assesses both the quality of the CIP work-process that leads to decision-making and the quality of the data/information available to the decision-maker to be used in the decision. Validation is distinctly different from prioritization. The CIP Validation process is designed to identify exactly where, for each project, improvements in decision-making confidence can be derived through improvements in process and/or data relevant to that given project. Once the project is deemed the right solution at the right time for the right cost – that is, its confidence level rating is sufficiently high to move forward with the project – the project can be prioritized (see Prioritization Tool).

The tool will be an Excel workbook that facilitates the analysis of feasible investment options for an asset or group of assets and assists in preparing a concise business case for the proposed decision. A Practitioner Review Team will review beta prototypes and modify as relevant. Two example case studies of up to six pages each – one each water and wastewater will be prepared in addition to developing user instructions. The final versions will be incorporated into SIMPLE.

Benefits

Capital investment is one of the most important – and one of the highest consequences – decision processes a utility faces. Because of the scale of the investment typically involved in a CIP, poor or wrong decisions will likely be very costly. High quality decisions only result from high quality work-processes coupled with high quality data. Decisions based on high quality work processes and high quality data have a high probability of generating those outcomes that were anticipated by the decision-maker – leading to higher confidence in the quality of the decision.

Interconnectivities/Interdependencies:

No direct.

3.23 Tool CIP 2 – CIP Prioritization

Priority: Phase 2 – Level 2

Description

Once the project list is validated, this tool assists in the prioritization of each project so that a specific CIP list can be adopted. The tool takes the form of a composite or master "go by" Excel spreadsheet of project attributes that might be considered as criteria for prioritization scoring; the practitioner may select appropriate criteria from the list and assign weights to each. A prioritization score is then calculated by the spreadsheet. It also incorporates Business Risk Exposure and Confidence Level Ratings to inform the prioritization process of embedded risks associated with a proposed project.

A Practitioner Review Team will review beta prototypes and modify as relevant. Two example case studies of up to six pages each – one each water and wastewater will be prepared in addition to developing user instructions. The final versions will be incorporated into SIMPLE.

Benefits

Due to funding limitations, not all valid projects may be budgeted for execution. Facilitating the determination of which of the validated projects get funded is the purpose of the prioritization tool.

Interconnectivities/Interdependencies:

Business Risk Exposure scores, Confidence Level Rating Scores.

3.24 Tool FE 2 – Funding Elements

Priority: Phase 3 – Level 2

Description

This tool walks the user through how to link investment decisions made through other tools with a financial spreadsheet to assess budget, rate, and, to a limited extent, financial condition impacts of those decisions. In the interest of cost reduction, this approach assumes that EPA's Cap Finance rate analysis tool would serve as the core "engine" for analysis. Such a linkage would facilitate "what if" scenario analysis as "optimized" decisions are traded off against feasible budget and rate strategies.

This tool will be an Excel workbook that facilitates the assessment of funding various levels of investment on future budget, rate, and financial condition using EPA's CAP Finance A Practitioner Review Team will review beta prototypes and modify as relevant. Two example case studies of up to six pages each – one each water and wastewater will be prepared in addition to developing user instructions. The final versions will be incorporated into SIMPLE.

Benefits

While funding an optimized set of strategies from a sustained performance perspective is the ideal, rarely does the policy board perceive that it can immediately raise rates to a full cost level. This Tool 1(assists the systematic search for an acceptable balancing of investment relative to budget and rate impact, and 2) provides a clear picture to staff and the policy board of the impact of the "ideal" life cycle based solution.

Interconnectivities/Interdependencies:

All other tools contribute to the context of applying this tool.

3.25 Renewal Cost Projection "Nessie Curve" Tool

Priority: Phase 3 – Level 1

Description

This tool facilitates a systematic depiction in graphic format of projected renewal costs over a defined period of time based on assumptions about remaining useful life and renewal costs related to existing assets. The tool can be used to depict cost patterns for an individual asset or a group of assets.

The tool will be an Excel workbook that facilitates a systematic depiction in graphic format of projected renewal costs over a defined period of time based on assumptions about remaining useful life and renewal costs related to existing assets scores as described above.A Practitioner Review Team will review beta prototypes and modify as relevant. Two example case studies of up to six pages each – one each water and wastewater will be prepared in addition to developing user instructions. The final versions will be incorporated into SIMPLE.

Benefits

Most utilities project capital costs only for at most a five year window. The life cycle of most utility type assets far exceeds that time frame. This means that both staff and the policy board have at best a minimal perspective of the larger renewal requirements picture. This can lead to short term decisions to defer investment due to short term issues that have dramatic long term implications. Depicting renewal requirements for a much longer time frame "tells the story" in much broader terms to those participating in the decision process. Such transparency tends to substantially improve the decision environment.

Interconnectivities/Interdependencies:

Condition, Major Failure Modes, Residual Life, Replacement Costing.

CHAPTER 4.0

IMPLEMENTING ASSET MANAGEMENT: CRITICAL SUCCESS FACTORS

GHD held a facilitated, one day workshop with the Working Group to identify inhibitors and success factors based on member experience in the creation and continuation of an AM program. The Working Group met at WERF headquarters on May 7-8, 2008 to identify and discuss inhibitors and success factors. The suggestions emanating from that workshop have been incorporated into this report. Note: the text presented below will be incorporated into SIMPLE in the form of revisions to existing SIMPLE text and may be somewhat modified to fit the SIMPLE narrative context.

4.1 Identifying Inhibitors and Success Factors

Most public agencies implement some form of asset management to move from a "design, construct, run and monitor" management regime to one that also focuses on long term planning for renewal, replacement and disposal of infrastructure assets. The WERF toolset for asset management implementation includes SAM-Gap which assesses seven (7) primary elements of asset management. The assessment is a first step to identifying organizational change that will be needed to move towards an asset management mode of operation.

The benefit from asset management implementation efforts depends on the quality of the change management process; recognizing what needs to be done, understanding the unique inhibitors to success, and implementing the key success factors throughout the change management process.

4.2 Inhibitors – A View From Inexperienced "Starters"

Facilitated work sessions were held with two groups of public agency participants: Group 1 being the less experienced "AM Starters", and Group 2 being the experienced "AM Veterans" in the SAM Challenge. The two groups identified some commonalities in initial success factors. When asked how an agency can get started in asset management, and what the agency should consider or be aware of to ensure a successful start, the AM Starters (Group 1) identified the success factors shown below in Table 4-1.

Group 1: (Less Experienced or "Starters" Group)

Table 4-1. Summary Success Factors (To Get Started).

Management commitment – management needs to understand that asset management is a process and not a project and has a long term focus. The approach to implementing asset management must be tailored to suit the individual utility.
Policy to implement asset management from governing body – asset management will only succeed where the governing body has made a commitment to provide funding support for the program
Defined benefits to stakeholders – the benefits from an asset management program must be clearly explained to all stakeholders, and ratepayers
Information technology systems that support knowledge management are essential ingredients in implementation of an asset management program
Adequate staff resources – staff should be solely dedicated to asset management implementation and it should not be added to their regular duties.
Training and education should take place at all levels in the organization
There must be adequate long term capital investment commitment by the governing body
Internal champions, who promote the benefits of asset management should be identified at all levels in the organization
Ways to measure progress towards implementation must be identified
The organization should have access to the various WERF asset management tools available through SIMPLE (Sustainable Infrastructure Management Program Learning Environment).
Phased Implementation – A realistic plan for implementation of asset management must be developed that takes into account resource availability and the unique circumstances at the utility. The plan should be phased with prioritized actions and must be flexible to accommodate changes as asset management knowledge improves over time.
A commitment must be made to continuous improvement in asset management so that decisions are made with an increasing level of confidence over time.

Experience shows that a structured approach to incorporation of the success factors helps with organizational change and long-term organizational learning. For example, establishing asset management policies early that embed asset management processes in the business planning and capital improvement planning cycles provides a monitored mechanism for change that often requires only an augmentation of existing processes. Continuous improvement in asset management is then kept alive by including asset management factors in process performance management and continuous improvement activities around the newly augmented business planning and capital improvement planning processes.

In addition, audit tools, like SAM-Gap and those that meet such international standards as the United Kingdom's PAS 55 standards, can be used to monitor asset management implementation. However, the organization may want to develop some internal monitoring methods to address the changing internal and external inhibitors to successful asset management implementation.

Tables 4-2 and 4-3 outline some common external and internal inhibitors to success. These were developed through a facilitated work session with SAM Challenge public agency participants. The group was asked to identify the key external and internal inhibitors or road blocks and how those inhibitors have been and might be successfully managed or mitigated.

Table 4-2. Inhibitors (External) to SAM Implementation.

Factor	Manage:	Mitigate Through:
Governing Body Indifference	♦ A mindset prevails that assets are performing satisfactorily so active management strategies are not required. ♦ The concepts and long term benefits associated with asset management are not understood or accepted by the governing body ♦ Political focus on short term initiatives funded from infrastructure budgets, such as leaving a new Police Station as a legacy	♦ Education of the governing body on the benefits of asset management ♦ A change in senior level management that is more supportive of asset management ♦ Benchmarking and interactions with peers at other organizations where asset management has been successfully implemented.
Financial Concerns ♦ Short term and immediate needs outweigh the long-term payoffs from asset management ♦ A perception by politicians and ratepayers that all rate increases should be either minimized or avoided	♦ Inadequate funding of long-term needs ♦ A perception that sources other than rate increases should be generated to fund infrastructure improvements ♦ A natural reticence by ratepayers to meet the cost of additional staff for an asset management program.	♦ Tradeoffs must be explained, such as funding infrastructure renewals rather than other competing initiatives ♦ Develop a long-term investment plan based on the 10 step asset management process ♦ Clearly articulate the benefits of long-term sustainable service delivery ♦ Engage with the community to explain the benefits and costs of asset management. Explain the likely consequences if inadequate long term investment is made. ♦ Explain the need to ensure inter-generational equity in meeting the costs of long term asset renewals to reassure ratepayers that they are only meeting their fair share of the costs. ♦ Graphically illustrate the long term funding needs through such mechanisms as a "Nessie" curve which shows the year by year costs to maintain sustainable services. ♦ Ensure that all capital improvements proceed only after the development of a robust business case which explicitly identifies the costs and benefits of the proposed work. ♦ Develop metrics to measure the confidence level for all asset management decisions so that continuous improvements can be clearly demonstrated to stakeholders to justify the long term investment plan.

<table>
<tr><th>Factor</th><th>Manage:</th><th>Mitigate Through:</th></tr>
<tr><td>Lack of understanding of asset management processes and inadequate measurement of outcomes</td><td>
<ul>
<li>The extent of understanding of processes will vary among utilities</li>
<li>It is challenging to demonstrate a positive outcome when large expenditures are involved with significant rate increases</li>
<li>Asset management is seen as a complex set of inter-related processes that are difficult to understand and rationally apply.</li>
<li>It is difficult to establish measurable performance standards that demonstrate the short and long term savings that will accrue from asset management.</li>
</ul></td><td>
<ul>
<li>Educate all stakeholders on the long-term benefits of asset management, as well as the levels of service consequences from not implementing asset management</li>
<li>Benchmark key performance indicators to demonstrate the positive benefits from asset management</li>
<li>Provide details of successful implementations of asset management at other organizations</li>
<li>Explain the role of business cases developed on a consistent basis in prioritizing expenditure needs</li>
<li>Provide actual case studies where an organization has introduced and then successfully implemented asset management</li>
</ul></td></tr>
<tr><td>Lack of Relevant Public Support</td><td>
<ul>
<li>An asset management promoter is required to enlist community support</li>
<li>Community expects consistency between water and wastewater</li>
<li>Resistance from unions to contracting out of services</li>
</ul></td><td>
<ul>
<li>Ensure that processes are transparent and easily understood by the community – particularly the setting of levels of service.</li>
<li>Explain the benefits at both a community and personal level</li>
<li>Articulate risks along with risk mitigation strategies that are easily understood</li>
<li>Address concerns about intergenerational equity of funding proposals</li>
<li>Develop genuine processes for engaging with and listening to the views of the community in setting levels of service and explaining funding implications of asset management.</li>
</ul></td></tr>
</table>

Table 4-3. Inhibitors (Internal) to SAM Implementation.

Limits on rates and revenues	♦ General moratoria on adding staff, or renewing assets	♦ Prioritize available funding through risk profiling (probability of failure, consequence of failure, risk mitigation) to optimize expenditure on critical assets. ♦ Accentuate the benefits of an asset management program to staff in making their jobs easier and more satisfying in the future. ♦ Encouraging staff to have a long term perspective rather than continuing in a reactive mode. ♦ Demonstrate the benefits of optimized investment decisions (maintenance, operations, capital renewal, capital augmentation) in delivering long term sustainable services at the lowest life cycle cost.
Lack of Management Commitment	♦ Variable level of support amongst senior management team	♦ Identify asset management promoters throughout the organization champions. ♦ Highlight benefits and positive outcomes from asset management. ♦ Stress the focus on long term planning to deliver sustainable services ♦ Impart asset management knowledge to management team ♦ Ensure there is an understanding that continuation of reactive approach will inevitably lead to declines in levels of service
Lack of awareness about relevance of asset management	♦ Need to educate a large group of people in order to move forward	♦ Explain that asset management is a paradigm shift and a new way of doing things to ensure long term service delivery at the lowest life cycle cost
No clear, common definition of asset management; no common strategic framework	♦ Differing views may exist within an organization of what asset management means	♦ Create project team to develop an asset management charter that clearly enunciates goals and objectives ♦ Set up structured training on asset management concepts for staff and management within the organization ♦ Make use of web-based educational material such as WERF's SAM GAP and SIMPLE
No perception of urgency	♦ Lack of prioritization or perception of urgency ♦ View of management that infrastructure is performing satisfactorily so there is no need to change the approach.	♦ Project full cost of maintaining services to sustain performance and show impact on current rates. ♦ Clearly demonstrate the impact of the funding gap for renewal of assets.
Inadequate IT systems integration	♦ Completion and full implementation of GIS spatial systems ♦ Standardization of core processes and use of CMMS in decision-making	♦ Prioritize integrations that will generate maximum benefits for the lowest initial outlay. ♦ Explain how the benefits of an asset management program might justify the IT integration expenditures.

		♦ Find ways of managing asset knowledge until integration is complete ♦ Prioritize system functionalities and focus on those delivering the greatest benefits
Levels of service expectations not well defined	♦ Level of service standards do not exist or are not measured	♦ Engage with the community to explain the inter-relationship between cost and levels of service ♦ Understand the difference between levels of service and performance measures ♦ Articulate current levels of service as a starting point for community engagement.
Limited staff resources devoted to asset management	♦ Lack of resources with the ability to devote time to asset management ♦ Lack of trained people with skills in asset management	♦ Obtain management commitment to allowing resources dedicated to asset management. ♦ Develop prioritized work plan for the training and skill-enhancement of staff matched to resource commitment ♦ Demonstrate genuine and realistic resource needs, identify benefits in support of business case to obtain required resources
Change Management	♦ Insufficient impetus to changing the status quo ♦ Implementation of change is perceived as being too much work ♦ The need for change has not been made clear	♦ Empower people to introduce change ♦ Include people in change management discussions by clearly articulating the need for change ♦ Prepare and implement a change management plan
Acceptance by staff	♦ Low levels of understanding of the benefits of asset management in comparison to negative aspects.	♦ Make sure the rewards from asset management are well explained ♦ Emphasize the importance of knowledge management in making jobs easier ♦ Follow through with implementation and clearly explain to staff the process for implementation
No clear driver for asset management	♦ No significant asset failure occurring recently to reinforce the need for improved asset management	♦ Show real benefits from case studies where other utilities have moved from a reactive to proactive regime. ♦ Encourage staff to visit peer agencies and discuss asset management experiences ♦ Develop "state of assets" report which portrays the state of all infrastructure assets
Perceived threats to staff autonomy	♦ Uncertainty and disagreement about where responsibility for where AM should be in the organization	♦ Focus on the need for the appropriate organizational structures to deliver sustainable services

4.3 Inhibitors – A View from Experienced "Veterans"

A group of SAM Challenge public agency participants that have at least several years of experience in asset management implementation were asked what the key inhibitors or challenges have been to sustaining a successful program. The resulting set of inhibitors, effects, mitigation and management activities are summarized in Table 4-4. The word "STOP" in the text box signifies those inhibitors that were felt might bring an existing asset management program to a halt.

Table 4-4. Inhibitors to SAM Implementation: A View from Experienced "Veterans".

Factor	Effects	Manage/Mitigate
♦ Change in senior management within an organization seeking to implement an asset management program	♦ Interrupts implementation process	♦ Ensure ongoing demonstration of value of program ♦ Maintain buy-in of entire team ♦ Establish an asset management steering committee to ensure ongoing commitment ♦ Embed new business processes and practices ♦ Undertake regular external reporting of value derived from asset management program
♦ Funding resource constraints	♦ Effort not sustained, variable commitment to program and schedule changes delaying implementation	♦ Reinforce business case – demonstrating value for money from asset management program ♦ Ensure there is an asset management champion in the leadership group who can advise policymakers on the consequences of funding constraints ♦ Demonstrate successes achieved throughout the program implementation ♦ Continually highlight long-term benefits from asset management and the potential consequences of not adopting asset management
♦ Resistance from operation and maintenance staff ♦ Why fix it when it is not broken and requires additional work ♦ Getting push back from staff trying to protect their long standing ways of doing things ♦ Sustaining support from staff during a long period of implementation ♦ Reluctance of stakeholders to meet increased rates used by staff to resist change	♦ <u>**STOP**</u>: Impact on asset operators who are key to service delivery ♦ Change in management or personnel may result ♦ leadership support may falter ♦ funding/ resource constraints	♦ Business case – demonstrate value for money and need to do things differently ♦ Champion/ in leadership/ gatekeeper ♦ Demonstrate how asset management will ultimately make jobs easier to perform ♦ Highlight long term benefits

♦ Inconsistent approach to asset management implementation	♦ <u>STOP</u>: Failed IT implementation and organization change management builds skepticism and wearies staff to absorb and embrace further change	♦ Business case for AM implementation and IT system development ♦ Follow through on implementation
♦ IT solutions not compatible	♦ Limits benefits of asset management implementation and potentially leads to disillusionment of staff	♦ Business case for asset management implementation and IT system development needs to be clearly articulated.
♦ Lack of understanding of asset management	♦ If potential benefits are not recognized at the outset there may be problems in the program getting off the ground	♦ Business case – demonstrated value for money ♦ Champion/ in leadership/ gatekeeper ♦ Demonstrate success as it occurs ♦ Highlight long term benefits
♦ Other commitments or existing workload	♦ Limits participation by staff in the implementation process	♦ Adjust pace of program to align with resource availability ♦ Ensure program is focused and prioritized to gain maximum efficiencies ♦ Prepare a detailed and realistic work plan ♦ Demonstrate value of program ♦ Promote the program as an "investment" as opposed to improvement ♦ Decentralize elements of program to overcome short term resource commitment issues
♦ Stakeholders/other departments do not see asset management goals and objectives as being important	♦ Limits benefits of program due to lack of commitment across organization.	♦ Use business case to demonstrate value for money ♦ Explain and obtain agreement to program goals and objectives ♦ Utilize asset management champion to convince detractors of benefits of asset management ♦ Highlight early successes resulting from program implementation ♦ Highlight long term benefits of program
♦ Question if asset management is adding value	♦ <u>STOP</u>: Economic downturn/drought could bring program to a halt ♦ Not universally accepted that asset	♦ Succession planning – identifying staff for the future with clearly defined career development paths ♦ Knowledge retention – by key individuals documenting their knowledge and role ♦ Diverse multi-function asset management delivery team with staff

Factor	Effects	Manage/Mitigate
	management adds value ♦ Retirement of key staff could lead to reduced ability to sustain program	representation from across organization ♦ EDMS – linked to CMMS (operations and management manuals) ♦ Ensure staff complete and maintain certification. ♦ Update job descriptions to clearly define roles and responsibilities

4.4 Summary of Success Factors

Consolidating the insights of the two groups, the following factors were deemed essential for ensuring a successful implementation:

- Identify quick wins – demonstrate value.
- Find an asset management champion – preferably two champions, one at the executive level and one at the mid or upper management level.
- Communicate (Internal/External) extensively and effectively
- Develop and manage expectations of stakeholders.
- Recognize the importance of training/skill building/career paths.
- Assertively seek buy-in at all levels of the organization through participation, communication, and education.
- Develop a cross-functional asset management steering team.
- Involve senior management from the start, and if time permits, involve staff from every level from the start, but be sensitive to and resolve early any miscommunications, misunderstandings, or conflicts arising.
- Develop a detailed work plan – include as much detail as possible so that any future changes can be accommodated.
- Emphasize the fact that asset management is a long-term (continuous) process that requires upfront and ongoing commitment to continuous improvement.
- Set short term and long term goals and manage associated expectations.
- Make sure that key people have the time and commitment to the asset management program.
- Devise strategies to overcome staff resistance and recognize that incremental implementation may be necessary to embed the program.
- Keep communication channels open at all levels – develop good communication plans at the outset.
- Develop and continuously tout business case for implementation of an asset management program.
- Carry out the SAM-GAP analysis early on and as a team to determine the current state of asset management in the organization.
- Network with other entities that have implemented asset management.

- Develop creative ways to implement asset management by tying it in with what people are already doing (e.g., data collection during maintenance activities).
- Educate and communicate with all stakeholders, but keep the messages simple, honest and straightforward.
- Petition state industry trade associations to promote asset management and encourage them to realize the benefits to members from developing asset management training programs.
- Implement the program incrementally and in manageable chunks.
- Develop an overall framework and specific plan to continuously move forward.

4.5 Financial Implications of SAM: Messages for Governing Bodies

In addition to technical and organizational factors that underlie implementation of SAM, the Participant Team identified a number of financial issues that need to be communicated clearly to governing bodies that oversee asset management budgets:

- In the past, infrastructure investment programs were determined primarily based on the age of an organization's assets. Now, asset function, condition, criticality, and operating environment must be considered. Minimizing asset life cycle costs results in savings that can be reinvested as renewal, replacement and new construction. Reallocation of investment is the primary saving achieved in new asset management implementations.
- Describe the state of current assets, and replacement and/or renewal costs in order to communicate the full financial and service level implications of ageing infrastructure and environmental pressures.
- Asset management can help make decisions to improve the budget by promoting an understanding of multiple funding scenarios and possibilities.
- Agencies can no longer afford to "repair by replacement". Infrastructure needs to be maintained with the right level of investment to sustain service levels and bond ratings. Asset management processes decide when and which assets to repair, replace or rehabilitate.
- Communities can leverage "Whole of Government" savings by adopting best asset management practices in all sectors that manage physical assets.
- Few things in a community are as important to long term well being as sustained performance of infrastructure; so it is incumbent upon asset managers to demonstrate a real business case for re-investment in infrastructure.
- What worked 30 years ago doesn't work today because the challenges are different. The water industry is realizing that planning for infrastructure is not just about design and construction – it's about managing the life cycle of assets from creation to disposal
- Maximize the use of existing assets
- Virtually all agencies are facing aging infrastructure with limited funding – so there is a need to carefully plan
- Indicate the risk of not doing asset management by highlighting business risk exposure, political risk, etc.

CHAPTER 5.0

ALTERNATIVE ROUTES TO IMPLEMENTATION OF ASSET MANAGEMENT

One task of the WERF Strategic Asset Management (SAM) project was to review, clarify, and modify the various pathways to implementing an asset management program (pathways) discussed in SIMPLE. In addition, a prototype staged guidance checklist that guides utility AM teams through implementation steps is to be developed, vetted through the Working Group, and added to SIMPLE.

The Working Group met at WERF headquarters on May 7-8, 2008 to review the implementation pathways in SIMPLE and to offer suggestions to clarify or strengthen the discussion in SIMPLE and to test a prototype of the Pathways Tool. This chapter presents suggestions emanating from that workshop. (Note: this material will be incorporated into SIMPLE in somewhat modified form.)

5.1 SIMPLE Review – "Options for Getting Started"

WERF's asset management knowledge base, SIMPLE, describes seven pathways that a utility might pursue to implement an asset management (AM) program (see Table 5-1). The most appropriate pathway for an organization will depend upon organizational and asset performance characteristics of the utility. Organizational attributes that are highly relevant, for example, include:

- Size of the organization and customer base
- Nature of stakeholder issues and interests ("green/environmental", lowest possible rates, growth oriented, predominance of legacy systems, etc.)
- State of repair of the infrastructure and financial health of the organization
- Corporate drivers of performance
- Management reporting structure and lines of accountability, culture and staff competencies

As discussed in Chapter 4.0, a number of factors affect the likely success of AM implementation, including the degree of management and governing board commitment, and availability of adequate IT systems and staff and financial support. In selecting an appropriate implementation pathway, the following commonly experienced key success factors should be considered:

- Management Commitment – implementation of Asset Management (AM) business practices is a process, not a project. It requires investment and unique solutions and approaches for each utility, including the participation of cross-functional teams and change management.
- Asset Management Policy – a successful AM program requires guiding policies adopted and driven by the governing body with linkage to the clear benefits to stakeholders.

- IT Systems support – development of the knowledge management elements required to support asset management decision making must go hand in hand with introducing AM business practices.
- Adequate Resources -staff and financial investment by the utility must be fully recognized and allocated to the program.
- Training and Education – all levels of the organization – governing body, staff, public – will require varying degrees of continuous training and education.
- Internal Champions – champions are required at all levels of the utility.
- Continuous Improvement – ways to measure progress, realistic plan for continuous improvement

It is not unusual for an AM champion to emerge through trial and error from within plant mid-level management staff. Often the champion achieves efficiency gains and performance improvements through the application of such practices as Reliability Centered Maintenance (RCM), Predictive Maintenance (PdM) to the operational environment or service level, and risk-based approaches to capital renewal planning. If benefits are achieved at the treatment plant level, for example, then other parts of the organization are likely to derive benefits from the same approach. A transfer of the associated strategies and procedures to other plants or maintenance activities can occur without a formal corporate policy on asset management, or executive level commitment to asset management. The extent to which this can occur depends on the organizational culture and the level of communication and information sharing between operations and maintenance staff, supervisors, and managing engineers across the organization.

However, maintenance, operations, and engineering are not always the areas where the organization can achieve the greatest benefit from asset management implementation. In many cases, the largest benefits can be realized at the board, commission, and executive management level, where the enabling decisions are made regarding revenue, funding strategies, capital investment, customer and stakeholder relationship management and project delivery and service strategies (e.g., public/private operations and maintenance contracts).

While implementation of asset management often starts in the engineering part of a utility, it usually spreads to other parts of an organization as a result of the need to keep pace with investment in the infrastructure through the peaks in the forecast replacement/renewal profile ("humps" in the "Nessie curve"). A corporate approach to asset management implementation results in a corporate level understanding of renewal liabilities and risk exposure. The organization is then better able to engage with customers and stakeholders to determine appropriate service levels and revenue and funding strategies to minimize rate shock, while maintaining the value or integrity of infrastructure assets for future generations.

A desirable goal for any organization is to exercise "good engineering" as well as achieve a strong corporate understanding of infrastructure funding needs and performance. The key question when selecting an appropriate asset management implementation pathway is: Where will the greatest benefits be achieved? A holistic view of the organization is generally necessary in order to make the right judgments and select the best approach.

Table 5-1. Summary of Common Pathways for SAM Implementation.

SIMPLE Pathway	Characteristics
1. Problem Asset or System	♦ Applies asset management principles to the management of a single problematic asset, asset group, or system ♦ Can deliver immediate benefits for problematic parts of a network or system
2. CIP Validation	♦ Applies asset management principles to investment decision-making and identifies projects that could be deferred, removed, or re-evaluated for inclusion in the Capital Improvement Program (CIP). Savings include: debt service on deferred and abandoned projects, improved use of capital, operations and maintenance costs ♦ Good for organizations that are becoming overwhelmed by growth and renewal capital projects, with high debt/equity ratios and increasingly less able to obtain financing ♦ Provides a sanity check on the size and content of the CIP through the application of a consistent, structured approach to evaluating projects using: confidence level ratings assessments, risk exposure scores, and lifecycle cost analysis ♦ Demonstrates the business case for applying asset management principles to the capital investment planning process
3. Existing AM Processes & Practices	♦ Pilot study to apply asset management principles to known process problem areas; risk assessment, optimized maintenance planning, capital investment validation, automation of operations activities, data management, optimized renewal decision making, condition assessment ♦ Good for organizations with robust processes and an understanding of where change can deliver the greatest benefits
4. Full Needs Analysis	♦ Good for establishing a clear corporate roadmap and business case ♦ Assessment of current performance against best practice – processes, practices, organization, people issues, information systems, data and knowledge ♦ May also involve a diagnostic on the operations and maintenance functions ♦ Identification of improvement projects and development of business cases ♦ Development of an implementation schedule/program
5. Information Systems	♦ Good for organizations considering implementing an Enterprise Resource Planning System, an Enterprise Asset Management System, or upgrading a maintenance management system ♦ Around 70% of system implementations fail because of inadequate change management and matching of the system configuration to the process support or functionality requirements of the organization ♦ Sets the foundation data and systems functionality in place for ongoing analysis and decision making – tactical (day to day operations and maintenance planning and execution) and strategic planning.
6. Modified Gap Analysis	♦ SAM Gap or a more sophisticated analysis (depending on the size of the organization and level of asset management capability) identifies top 10 "quick win" projects. Business cases are prepared for those projects and implementation proceeds. ♦ Doesn't require a full level of competence or understanding across all silos before consensus to proceed is obtained
7. Step-by-Step AM Plan	♦ Produces a basic Asset Management Plan (AMP) for the entire asset portfolio ♦ Good for all organizations ♦ Brings together a tangible deliverable from the asset management process ♦ Engages all levels of the organization, including elected representatives, customers and stakeholders (if desired)

For example, in organizations where a specific major process (for example, a particular aeration basin) or a specific type of asset (for example, unlined cast iron pipe) poses a significant challenge, Pathway 1 – Problem Asset or System Approach may well be the appropriate approach. Pathway 1 begins by focusing on a specific set of asset management practices and techniques (for example, condition assessment, failure mode analysis, or residual life analysis) or on a selected asset, class of assets, or major process in what might be termed a pilot mode. The practices and techniques would be applied by the team of employees who have the aggregate responsibility for investment decision-making over the life of that asset (for example, engineering, operations, maintenance) and related support personnel who play a role in the larger business process (for example, IT and Budget). The pilot would be used to test, then adapt, generic practices to the actual operating environment of the utility – usually in the form of written protocols. The developed practices then would spread through the organization, ideally moving both up and down the organizational structure.

The process of continuous improvement will enhance asset management planning by identifying business processes – such as determining long term renewal "annuity funding levels" and funding strategy analysis – that can be improved to begin the process of actively optimizing lifecycle cost. Within the context of an adequate or enabling income generation model, lifecycle costs can be optimized by applying the already well developed operations, maintenance and capital renewal project identification processes.

The seven implementation pathways are built around two approaches:

1. Identify changes required to deliver the greatest overall benefit to the organization – SAM Gap (a gap analysis tool used to identify the "as is" and "to be" states), or a more comprehensive assessment of the utility's asset management practices and procedures, and the development of a prioritized implementation plan (see Figure 5-1).
2. Initiate projects that will generate a quick return on investment (known as "quick wins").

As one size does not fit all, combinations of the pathways are often adopted as a way forward. The challenge is to identify where the greatest benefits will be achieved and to structure a way forward that navigates the unique cultural, organizational, resource, and asset issues that an organization faces. Chapter 4.0 discusses factors that often inhibit implementation of AM in an organization, including a lack of common understanding of AM and its benefits and insufficient resources devoted to AM. Successful implementation of AM may require education, including benchmarking against other organizations, change management approaches (development of financial assessments and the business case), and outreach to stakeholders.

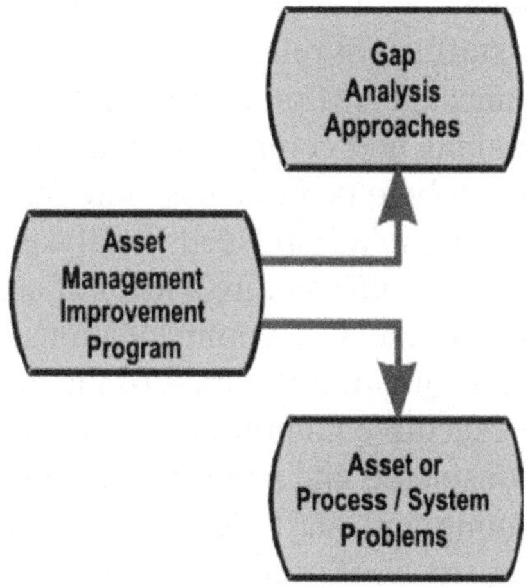

Figure 5-1. Two Major Pathways to Improve Asset Management.

5.2 Choosing an Implementation Pathway

Tables 5-3 through 5-5 at the end of this section summarize a selection of the organizational and asset cost and performance factors that can be utilized to characterize the asset management competency of an organization. By considering each of these factors in terms of the inhibitors identified above, and the gap between what the existing and desired states, it is possible to get a sense of the level of effort required as well as the appropriate pathway(s) to move the organization in the desired direction.

To use the pathways checklist (Table 5-5), users should select the boxes in Tables 5-2 and 5-3 that best describe their organization. The selected boxes should then be compared to the selection criteria in Table 5-5, which summarizes the mix of organization and asset cost and performance characteristics that each of the pathways is most able to address. The level of benefit for each pathway varies according to the starting point for the organization.

Hypothetical examples, based on a series of assumptions are set out below to illustrate the use of the checklist to select the appropriate pathway(s).

5.2.1 Example 1

Checklist Factors:

Organization A serves a population of 100,000 persons or more and the average age of the infrastructure is 50 years. The asset management initiative is being driven at the engineering or operations level. There has been a chronic underinvestment in infrastructure assets. The infrastructure system is assessed as being in a borderline state of repair and the financial situation is favorable with respect to bond rating and scope for increasing rates and charges. The organization has experienced some sanitary sewer overflow (SSO) violations. Planned/ preventive maintenance processes are in place and engineering staff recognize the need for risk based strategic planning in order to keep ahead of regulations and aging infrastructure challenges.

Pathway Selection:

Given that the need for asset management implementation is being driven by engineering and operations level staff, Pathway 4, Full Needs Analysis could be premature for this organization. However, instances where the level of stewardship demonstrated by the engineering, maintenance/operations staff is not reflected in the level of investment in infrastructure usually indicates disconnected AM processes and the need for some improvements in both capital investment decision making and communications between engineering and executive decision makers. There are likely to be other systemic management issues across the various organization functions. Investment in a Full Needs Analysis would require involvement from staff across the organization, which would require a level of commitment from executive management that Organization A does not have. Recommendations coming out of a Full Needs Analysis would probably be far reaching and involve areas of the organization outside of engineering and operations. Without executive commitment, implementation of these recommendations may not even be considered and payback on the investment may never be realized. Also, without executive commitment, staff participation may not be at the level required to enable a complete and realistic assessment of needs.

The organization serves 100,000 persons, so is considered to be medium sized. If the culture is supportive, it is probably capable of leveraging significant benefits by demonstrating value at the engineering/operations level. The engineering/operations group may be successful in selling this to executive management with a business case for what is needed from other parts of the organization to achieve greater results.

The organization has not yet reached a point where it collectively agrees that a new path forward is required for managing its aging assets in the face of increasing regulatory requirements. If the engineering and operations divisions have the capacity to fund some AM implementation within existing budget allocations, and they also have the required motivation, a combination of pathways could be considered.

Pathway 1 – Problem Asset or System Approach and/or Pathway 7 Step-by-Step AM Plan Approach (the review of problematic assets/systems and the development of asset or system asset management plans) will generate new management information and datasets that will assist executive management decision making. The operations, maintenance and engineering group will be able to demonstrate funding needs packaged in the context of investment, risk, timing, and service levels. This could be particularly helpful in communicating the organization's needs for future investments and changes in processes and procedures to prevent future SSO violations, for example. This may create an impetus for change in other areas of the organization - to continually improve the generation of sufficient income and funding options, asset performance and lifecycle costs.

Organization A could couple this effort with Pathway 6: Modified Gap Analysis, limited to the assets under consideration. The Modified Gap Analysis would identify the business case for what the engineering and operations group might need from other parts of the organization in terms of support, such as: changes to the Capital Improvement Program selection process, information systems functionality and managerial accounting enhancements.

5.2.2 Example 2

Checklist Factors:

Organization B serves a population of over 500,000 persons and the average age of the infrastructure is 50 years. An asset management implementation initiative already has full commitment from executive management. Assets are in a borderline state of repair, and the financial situation is favorable with respect to bond rating and scope for modest rate increase.

There is limited use of AM processes for strategic planning and planned/preventive maintenance. Some AM data is collected and quality assured. Response times and service standards are below industry benchmarks. CMMS, Finance, GIS systems are functional, but not fully utilized to support AM. The organization is customer focused with distributed accountabilities for cost and service level performance. Staff and financial resources are available for AM implementation. No permit violations have occurred and energy and Green House Gas (GHG) emissions are within normal benchmarks. Initial discussions with engineering and operations staff indicate skepticism and a lack of change readiness.

Pathway Selection:

The fact that the infrastructure is in a borderline state of repair and the organization has not taken advantage of a healthy financial situation that could have enabled additional investment, coupled with the lack of strategic planning and the reactive maintenance environment, highlights a number of issues:

- Current management processes and practices are unlikely to be adequate for managing increasing probabilities of failure and risk exposure – the organization is probably already operating under high levels of risk and has likely been very fortunate to date with respect to violations.
- Significant change management initiatives will be required to bring the organization to a common starting point for moving forward with widespread asset management business transformation.
- Information systems are in place, but some re-configuration and change management work will be required to fully utilize their capabilities within a proactive maintenance and asset management environment.

Organization B is a candidate for Pathway 4: Full Needs Analysis, Pathway 6: Modified Gap Analysis, and Pathway 7: Step-by-Step AM Plan Approach, and Pathway 1: Problem Asset or System Approach. A combination of these approaches could be considered.

Since there is full commitment to asset management implementation from the executive management and there are a myriad of issues to address across the organization, Pathway 4: Full Needs Analysis could be considered. The Full Needs Analysis will provide a roadmap of the way forward with prioritized projects based on cost/benefit and payback period. The projects can also be prioritized based on the need to navigate the gamut of cultural, communications and political challenges. Pathway 6: Modified Gap Analysis could also be chosen if the change readiness of the organization will not support the level of participation that the Full Needs Analysis might require to achieve best results.

The organization might also benefit from Pathway 7: Step-by-Step AM Plan Approach to establish the process of developing a network asset management plan early in the implementation phase. The development, adoption, use, and regular revision of a network asset management plan is usually one of the most important change management activities to undertake. The network asset management plan becomes the control, or the vehicle for directing further asset management implementation initiatives by keeping the implementation process alive and focused.

Pathway 1: Problem Asset or System Approach can be used as a change management tool for demonstrating the value of the process to the engineering, maintenance and operations staff.

5.2.3 Example 3

Checklist Factors:

Organization C serves a population of 15,000 persons. The organization is customer focused with clear accountabilities assigned for both cost and performance. The infrastructure suffers from chronic underinvestment and is in a borderline to poor state of repair. There are limited staff and financial resources. The organization is starting to experience the classic problems associated with an aging and poorly maintained infrastructure system. The treatment plant requires a major overhaul to meet work place health and safety standards, and is in need of an upgrade to meet U.S. EPA requirements to avoid further drinking water violations. Distribution network crews are experiencing increasing problems with tuberculation, an inability to find buried valves to isolate the system to repair pipeline breaks, and the need to innovate to get around the lack of available spares to effect system repairs.

There is limited strategic planning or planned/preventive maintenance, and limited information systems tools. The directive for change is coming from the Executive Director. He is being approached by private system operators interested in engaging in a contract to operate and maintain the system, and he needs to evaluate the options and present them to the elected representatives.

Pathway Selection:

Pathway 7: Step-by-Step AM Plan Approach is probably the executive director's best option. The Step-by-Step AM Plan Approach will give him an asset management plan for his system that will include the capital, maintenance and operations cash-flow requirements moving forward. It will also give him an understanding of other investment needs such as the implementation of routine maintenance procedures, a program to identify where valves are in the field, and methods for determining the appropriate spares inventory.

The asset management plan can be used as a vehicle for consultation with the community regarding the upcoming cost for service and form part of the decision making process regarding private operation. If the organization and its stakeholders decide that private operation is not the most desirable option, the roadmap developed in the AM Plan can be used to move forward with the projects and funding strategy identified to improve the level of management of the system.

Table 5-2. Implementation Pathways Checklist.

Organizational Factors			
Population Served	<25,000	>100,000	>500,000
Organization Complexity	Asset & customer focused, clear accountability for cost and service performance	Customer focused, clear accountability for cost and service performance	Customer focused, distributed accountabilities for cost and service performance
Strength of Leadership	Directive of CEO/GM/Elected Reps	Limited corporate commitment however multi-silo initiatives in place	Operations level initiative/buy-in only
Change Readiness	Skepticism and resistance, executive is not change ready	Executive is change ready, engineering/operations not change ready	Executive is Change Ready, Engineering/Operations & Maintenance change ready
Priority/Schedule	Corporate Priority	Departmental Priority	Facility Level Priority
Resource Availability	Staff and financial resources available	Staff resources available, financial resources limited	Limited staff and financial resources
Process Robustness	AM processes connected across finance, engineering & operations functions	Planned/preventive maintenance processes in place – some use of AM decision making for strategic planning.	Limited use of AM processes for strategic planning & planned/preventive maintenance
Data Robustness	AM Data collected and quality assured	Some AM data collected and quality assured	Data collection haphazard, limited consistency
Information System Robustness	CMMS, Finance, CIS, GIS integrated with asset level analysis	CMMS, Finance, CIS, GIS utilized but not fully functional for AM	Limited CMMS, Finance, CIS, GIS, limited asset level analysis

Table 5-3. Performance Factors.

Asset Cost and Performance Factors			
Average Age of Infrastructure	> 30 years	> 50 years	> 75 years
Operating Cost Reductions	Operating cost benchmarks indicate that significant cost performance improvements can be derived through productivity improvements and asset lifecycle cost improvements	Operating cost benchmarks indicate that some cost performance improvements can be derived through asset lifecycle cost improvements	Operating cost benchmarks indicate that the operating cost structure is close to optimal
Capital Program – Funding Challenges	Assets are in a good state of repair, maximum reserve funds accrued, debt/equity, bond ratings favorable, and rate increase capacity available.	Assets are in a borderline state of repair, financial situation favorable – bond rating, rate increases meet requirements.	Chronic underinvestment in assets, limited ability to increase rates and charges, bond rating favorable
Environmental Performance	Clean Air/Water act/permit violations, or high energy consumption, odor/ GHG emissions	No permit violations. Energy/odor/GHG emissions can be improved	No permit violations. Energy, odor, GHG emissions within normal benchmarks
Customer Service	Customer service standards consistently reviewed and met	Response times and service standards acceptable	Response times and service standards below industry standard benchmarks

Factors shown in bold in Table 5-4 indicate those factors that are essential or strongly indicate that the pathway should be considered. Presence of additional factors increases the relevance or strengthens the pathway choice.

Table 5-4. Pathways Selection.

Organizational Factors
Asset Costs and Performance Factors

SIMPLE Pathways	Factors							
1. Asset/System Approach	**Operating cost benchmarks indicate that significant cost performance improvements can be derived through productivity improvements and asset lifecycle cost improvements**	**Operating cost benchmarks indicate that some cost performance improvements can be derived through asset lifecycle cost improvements**	**Assets are in a borderline state of repair, financial situation favorable – bond rating, rate increases meet requirements.**	Chronic under investment in assets, limited time horizon for price path coverage	**Clean Air/Water act / permit violations, or high energy consumption odor/ GHG emissions**	No permit violations. Energy/odor/GHG emissions can be improved	Response times and service standards below industry standard benchmarks	Facility Level Priority
	Limited staff and financial resources	Staff resources available, financial resources limited	Planned/preventive maintenance processes in place – some use of AM decision making for strategic planning.	**Limited use of AM processes for strategic planning & planned/ preventive maintenance**				
2. CIP Validation	Assets are in a borderline state of repair, financial situation favorable – bond rating, rate increases meet requirements.	**Chronic underinvestment in assets, limited time horizon for price path coverage**	Directive of CEO/GM/Elected Reps	Limited corporate commitment however multi-silo initiatives in place	**Skepticism and resistance among executive management**	Executive management is change ready, engineering & operations are skeptical/resistant		
3. Existing Process & Practice Approach	Response times and service standards below industry standard benchmarks	No permit violations. Energy/odor/GHG emissions can be improved	Assets are in a good state of repair, maximum reserve funds accrued, debt/equity, bond ratings favorable, and rate increase capacity available.	Assets are in a borderline state of repair, financial situation favorable – bond rating, rate increases meet requirements.	Response times and service standards below industry standard benchmarks	Planned, preventive maintenance processes in place – some use of AM decision making for strategic planning.	**Limited use of AM processes for strategic planning & planned/ preventive maintenance**	

4. Full Needs Analysis Approach	Assets are in a borderline state of repair, financial situation favorable – bond rating, rate increases meet requirements.	Chronic underinvestment in assets, limited ability to increase rates and charges, bond rating favorable	Customer focused, distributed accountabilities for cost and service performance	**>500,000 population served** **>100,000 population served**	**Directive of CEO/GM/ Elected Reps**	**Staff and financial resources available**	Planned/ preventive maintenance processes in place – some use of AM decision making for strategic planning.	**Executive is Change Ready, Engi-neering, O & M change ready**
5. Information Systems Approach	**Operating cost benchmarks indicate that significant cost performance improvements can be derived through productivity improvements and asset lifecycle cost improvements**	Operating cost benchmarks indicate that some cost performance improvements can be derived through asset lifecycle cost improvements	**Limited CMMS, Finance, CIS, GIS, limited asset level analysis**	Some AM data collected and quality assured	Data collection haphazard, limited consistency			
6. Modified Gap Analysis Approach	Response times and service standards below industry standard benchmarks	Clean Air/Water act/ permit violations, or high energy consumption, odor/ GHG emissions	Assets are in a borderline state of repair, financial situation favorable – bond rating, rate increases meet requirements.	Chronic underinvestme nt in assets, limited time horizon for price path coverage	Limited corporate commitment however multi-silo initiatives in place	Operations level initiative/buy-in only		
7. Step by Step Approach	All factors are supported by this pathway.							

WERF

CHAPTER 6.0

INFORMATION REQUIREMENTS FOR ASSET MANAGEMENT

The information requirements for asset management are described for each of the steps in developing an enterprise-wide Asset Management Plan. A process flow chart is used to highlight flows of information among assessment, analytical, and decision-making tasks. The process flow chart along with attendant descriptive text will be added to the SIMPLE knowledge base.

6.1 Asset Information Needs of Management

It is vital for owners of assets with large, complex portfolios to have a clear picture of their business performance. At the strategic level, each asset owner needs to be able to identify how well performance requirements (as determined by customers and stakeholders) are being met, and at what cost, and what will be required to continue to meet those performance levels well into the future. At the operational level, owners face important investment decisions on a daily basis, including:

- Should this asset be "run to failure"?
- Should an aggressive proactive maintenance strategy be pursued?
- Is it time to transition from a maintenance strategy to a capital renewal (repair, refurbish, replace) investment strategy? If so, which renewal strategy will be most cost effective?

The asset management paradigm requires the consideration of the entire life cycle process (as represented in Figure 6-1) in any asset investment decision. All elements of the life cycle need to be managed—from asset planning through the creation/acquisition phase into operations, maintenance and renewal and through to disposal.

Life cycle management can be relatively easy for a single asset, but when organizations own a mature network or portfolio that consists of hundreds of thousands of individual assets, the task becomes far more challenging. It is therefore important to understand where assets are in their life cycle so that their cost and performance can be effectively managed.

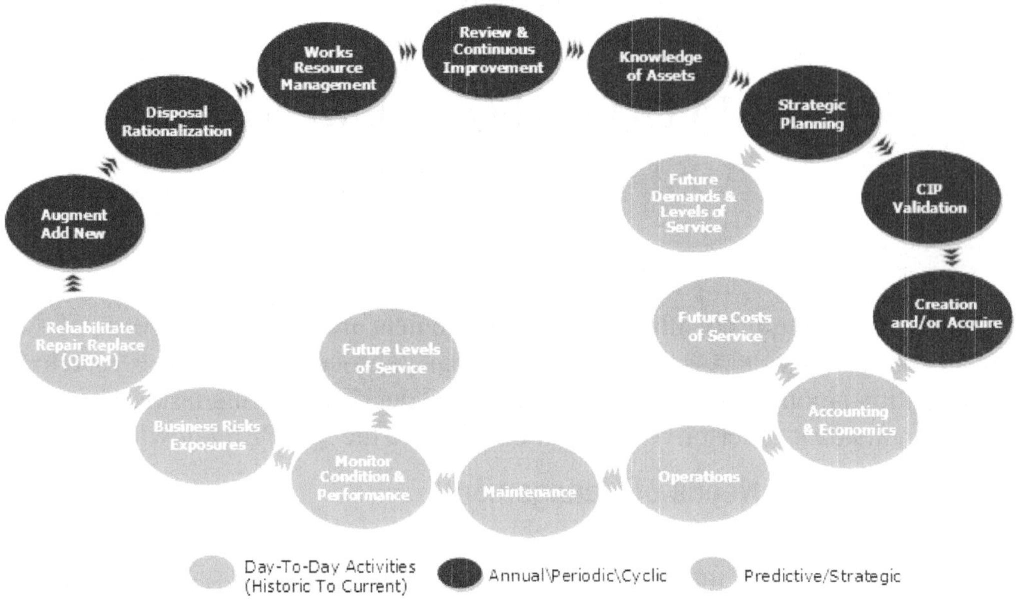

Figure 6-1. Asset Life Cycle – Extended Version.

6.2 The Importance of "Information"

Appropriate and readily available information is key to managing the life cycle of assets, which encompasses all activities from the initial identification of a need for a service through the planning stages to the actual delivery of the service provided by the assets.

Although the activities of life cycle management are generally carried out in sequence as shown in Figure 6-1, the flow of information about the assets takes place across various activities. Effective asset management decision-making requires analysis of information from multiple stages of the life cycle. For example, to carry out effective maintenance, relevant asset information is needed from the Strategic Planning and the Knowledge of Assets activities.

It has been said about asset management that the more we know about our assets, the better we can manage them. But collecting data – more importantly, building knowledge – can be prohibitively expensive. Thus, asset managers need to distinguish between information that is vital for effective decision-making and information that is helpful, but not essential to make effective asset investment decisions.

It should be noted that data is not information. Individual data points represent observations at a specific location and time. Information, on the other hand, is an understanding of the relationships between pieces of data; information is data that have been organized in temporal or spatial terms. Information contains structure, but little context.

Just as data is not information, information is not knowledge. Knowledge is often defined as "interpreted information" – information with a context, containing discernable patterns, where those patterns are interpreted within a relevant conceptual framework. In this simplified perspective, data are gathered, organized into information, and then the information is interpreted within a specific context and conceptual framework relevant to a specific decision. As knowledge, patterns of data eventually become a basis for prediction, and provide guidance towards a determined course of action.

6.3 What Information is Necessary for Asset Management Decisions?

Collecting data and implementing major data management systems requires considerable investment. Prior to a data collection effort for asset managers, therefore, a framework must be constructed that guides the investment decisions about what data to collect, and how to analyze and report them. Decisions about what data/information to collect must be driven by the questions the management team should be asking and what decisions they are facing. A determination about what data/ information is needed must be driven by a framework. The following discussion outlines an asset management information framework.

6.4 The Principles of Asset Management

The principles of asset management apply to all asset classes, such as passive assets (buildings and infrastructure), static assets (land), and active assets (mechanical, electrical, electronic, mobile plant, office equipment, software, etc). It is only the specific techniques, practices and life spans that will differ depending on the individual assets of interest.

What this means to most organizations, even those managing a broad cross-section of asset types, is that many of the principles of asset management are common to the organization as a whole. Consequently, a uniform set of principles can be applied to all asset groups allowing senior management to judge the relative merits of each activity or service. It is essential that asset management systems be able to provide the outputs required across the asset life cycle.

These principles are briefly characterized as:

1. **The "Value Added/Level of Service" Principle** – assets exist to deliver services and goods that are valued by the customer/stakeholder; for each consumer/stakeholder there is a minimum level of service below which a given service is not perceived as adding value.
2. **The "Life Cycle" Principle** – all assets pass through a discernable life cycle, the understanding of which enhances appropriate management.
3. **The "Failure" Principle** – time, usage, and the operating environment work to break down all assets; failure occurs when an asset cannot do what is required by the user in its operating environment.
4. **The "Failure Modes" Principle** – not all assets fail in the same way.
5. **The "Probability" Principle** – not all assets of the same age fail at the same time.
6. **The "Consequence" Principle** – not all failures have the same consequences.
7. **The "Total Cost of Ownership" Principle** – there exists a minimum optimal investment over the life cycle of an asset that best balances performance and cost, given a target level of service and a designated level of risk.

The challenge is to apply these principles in a systematic manner within a complex organization – through an asset management program.

6.5 Information and the Asset Management Program

It is important to consider the implications of the principles and concepts of asset management in order to adequately plan for the information systems needed for asset life cycle decision-making.

The basic elements of asset management can be organized into a business value chain that includes seven primary quality elements:

1. **Processes and Practices** – the methods and business processes used by an organization in managing its assets, such as providing for, maintaining and operating its assets and resources to deliver the expected service levels from the assets
2. **Data and Knowledge** – factors such as the quality, accuracy, and availability of asset related data and information
3. **Information Systems** – issues such as the functionality, level of integration and software used to support asset decision-making
4. **Service Delivery** – using the most appropriate market-tested means of providing an asset related service
5. **Organizational Issues** – refers to whether factors such as organizational structure are appropriate to deliver asset related services
6. **People Issues** – whether the staff supplying asset related services have adequate experience, training and support
7. **Asset Management Plans -** incorporating all the above processes into a planning cycle and producing the requisite planning documents

The above seven elements of the asset management value chain are represented in Figure 6-2. As indicated in the figure, information systems combined with data and knowledge play key roles in delivering an integrated asset management program.

Figure 6-2. The Asset Management Value Chain.

6.6 Specific Asset Knowledge

Successful management can be said to start with asking the right questions, then obtaining the information necessary to answer those questions. There are five core questions about assets that all infrastructure managers should be constantly posing to their management teams. These questions are:

Question 1: What is the current state of my assets?

- What do I own?
- Where is it?
- What condition is it in?
- What is its remaining useful life?
- What is its remaining economic value?

Question 2: What is my required level of service (LOS)?

- What is the demand for services by stakeholders?
- What do regulators require?

- What is my actual performance?

Question 3: Which assets are critical to sustained performance?

- How does it fail? How can it fail?
- What is the likelihood of failure?
- What does it cost to repair?
- What are the consequences of failure?

Question 4: What are my best O&M and CIP investment strategies?

- What alternative management options exist?
- Which are the most feasible for my organization?

Question 5: What is my best long-term funding strategy?

To address these questions, a manager needs not only accurate and timely information, but more importantly, the right information. Many organizations have bits of this information spread across disparate systems in the organization. The information often resides in legacy systems designed and maintained by separate organizational silos primarily for the use of staff within that department of the organization. The intent of an enterprise asset management system is to bring all of the relevant information together in a way that facilitates efficient life cycle investment decision-making for the entire asset portfolio across the whole organization.

6.7 A Decision Framework for Asset Management Information

In effective organizational management, knowledge and strategy go together. A knowledge-based strategy for a utility should focus on addressing the fundamental question - "what knowledge is required to sustain system and service performance over the long term at the lowest cost to our customers and stakeholders?" The following questions then assist in organizing the pursuit of "the right strategic knowledge":

- What is the organization's mission and strategies for achieving it?
- How do assets relate to the mission?
- What asset knowledge is required to achieve the organization's mission over the long run?
- Who in the organization needs the information?
- How will that knowledge be managed?

In the public sector, this knowledge is crucial to the budgeting process, where investment decisions about operations, maintenance, and renewal are rendered. Millions of dollars are at stake, so a poor decision based on incomplete and inadequate information has substantial life cycle cost ramifications.

6.8 Asset Management Information System Functionalities

In order to maximize the quality of investment decisions, it is necessary to understand the interrelationships and interactions among life cycle activities. The minimum necessary life cycle based functionalities for an effective asset management information system include:

Strategic Planning Functionality

- Asset register and detailed physical attributes
- Failure prediction
- Condition assessment
- Risk assessment
- Optimized renewal decision-making
- Budget rationalization and implications
- Life cycle costing

Maintenance Management Functionality

- Maintenance management
- Works/resource management
- Inventory management
- Spatial information
- Customer service/levels of service

Other Functionalities

- Creation, rehabilitation, and disposal (Capital Improvement Program)
- Operation and performance monitoring
- Valuation and depreciation

Beyond these minimum functionalities, an enterprise level asset management system may require additional information and assessment capabilities. Identification and description of more advanced functionalities and associated information requirements should be the focus of further WERF Strategic Asset Management research.

6.9 Information Flows

The asset management process starts with the development of an Asset Register and an understanding of the tie between the asset portfolio and Levels of Service and Customer Expectations. Based on the levels of service required from the assets, condition and performance can then be understood. These strategic planning activities will identify the right blend between capital and maintenance expenditures to ensure that the asset portfolio will deliver the required level of service in a sustainable manner.

Failure Prediction, Condition Assessment, Risk Assessment, and Optimal Renewal modules are structured such that basic, intermediate, and advanced techniques are available, depending upon the criticality of the assets being managed. For example, an asset that requires high levels of expenditure and presents a high risk to the organization may warrant the use of advanced techniques whereas inexpensive assets with a low consequence of failure may warrant only the use of the basic condition assessment techniques.

Consideration should include:

- Alternative failure modes;

- Risk the asset presents to the organization; and
- The options for treatment and management of the asset.

The works program resulting from the portfolio assessment advances through the **Budget Rationalization** process to match available funds with desired works and projects. Once rationalized, the **Capital Improvement Plan (CIP)** and **Operations and Maintenance** decisions drive the focus of the asset management plan.

Maintained assets are fed through to the **Operations and Maintenance** system, for cyclic or predictive based task identification. Assets to be acquired renewed or disposed are fed through the CIP update planning process, ensuring the asset register is updated accordingly (see Figure 6-3).

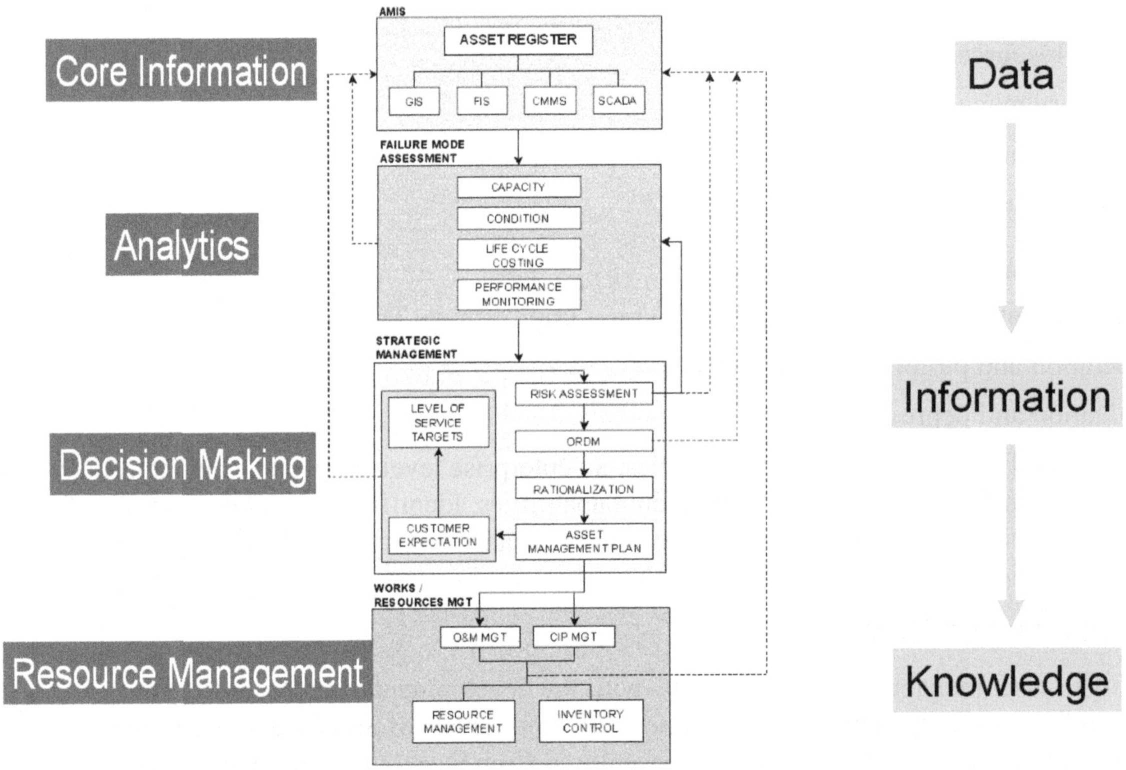

Figure 6-3. Asset Management Information Structure and Flow.

In the execution of work, the Resource Management module aims to match the available resources to the work required. This combines external and internal resources and identifies when tasks will be undertaken based on a priority allocation. The Resource Management module also captures costs against assets and activities for use in performance monitoring and downloading to the financial and maintenance systems.

The **Life Cycle Costing** module is used to aggregate information regarding historic and future asset costs, to allow the user to compare options (e.g., lease vs. own). This Life Cycle Costing information is a vital input into the Optimal Renewal module.

The **Performance Monitoring** module is used to assess performance of assets by capturing costing, condition, performance and other asset related activities. Non-performing assets that are identified are referred back to the strategic planning process for re-evaluation of the best treatment option (e.g., repair or replace options).

The **Customer Module** is used to track CIP requests and to ensure that issues raised by customers are handled in a business-like manner, by ensuring adequate and timely responses to service requests. This module can apply to both internal and external stakeholders.

Beyond the basic Asset Management information management described herein, a more comprehensive model for information and system functionality might be developed as an organization gains experience with AM. An example of a more detailed information management model with additional data feedback loops is provided in Appendix C.

6.10 Roles and Responsibilities

It is important to define roles and responsibilities for different activities in each module in the information flow chain. The likelihood of asset failure and poor management decisions increases where responsibilities are not defined, poorly defined, or where multiple groups have responsibility for an asset. Setting and rationalizing responsibility for assets should provide clear direction on responsibility to management.

6.11 Inputs and Outputs

Defining the inputs and outputs of the decision framework is also important where various divisions of an organization have responsibility for providing service or data to other divisions of an organization. For example, if one division is responsible for the condition assessment of an asset, but another division is responsible for the subsequent failure analysis, the exchange of information becomes critical to decision-making. It is therefore necessary for each group to know what inputs and outputs they produce that support the other divisions of the organization and vice versa.

6.12 Using the 10-Step Process to Define Core Information Requirements

Asset management boils down to a systematic resolution of the five questions posed in Section 6.6 by determining how best to answer the questions and how to effectively "tell the story" embedded in the answers to the questions. International experience over the past two decades points to the development of an asset management plan as key to both answering the questions and telling the story. An Asset Management Plan systematically:

- Characterizes the state of the assets;
- Identifies levels of service expected from the assets;

- Identifies "critical" assets (assets with both high probability and high consequence of failure);
- Identifies a set of the most cost effective maintenance, operations, and capital investment strategies based on the above; and
- Defines a funding strategy.

Typically, the plan also identifies what practices within the organization need to be strengthened and how the organization should go about such strengthening. An asset management plan may be generated at a single asset level (motor, pipe segment), a facility or subsystem level (a plant, a lift station, a class of pipe), a network level (water, wastewater, communications, etc.), or at an enterprise level.

The preparation and periodic review of asset management plans should provide:

- Consolidated technical and financial information relevant to the assets providing a service to defined groups of customers;
- A consolidated statement of current policies, strategies and programs adopted for the ongoing provision of these services;
- An understanding of the present and future demands on the assets;
- A current estimate of the long-term financial commitments necessary to maintain both the assets and the services they provide;
- A current evaluation of business risks associated with the failure of the assets;
- A summary of appropriate strategies to address shorter term operations and maintenance and longer term strategic planning of the assets;
- An information source that is capable of spanning organizational changes and the transfer of responsibilities between successive asset managers;
- A tool for the administration of contracts for operations, maintenance and renewals;
- A clearly established link between business objectives and asset management outputs;
- An ability to model future costs and asset performance; and
- An ability to identify future work and staff needs.

The process of building an Asset Management Plan reduces to 10 steps, depicted in Figure 6-4, that are directly related to the five core management questions (Section 6.6) and are carefully built around the seven asset management principles presented in Section 6.4. It should be noted that certain "best practice" processes and specific information are necessary for the execution of each of these steps. To successfully execute the steps, an organization must master the basics of the associated practices and generate the relevant information.

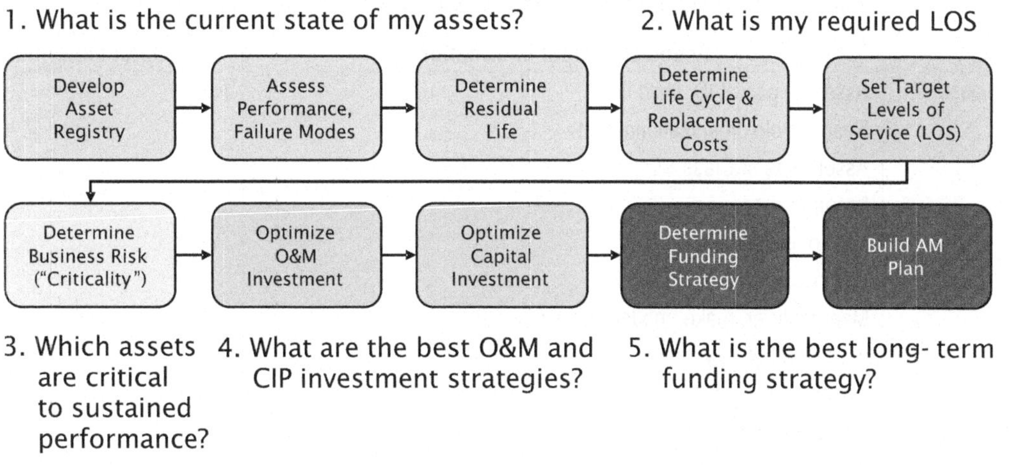

Figure 6-4. 10-Step Process for Developing an Asset Management Plan.

This 10-step model forms the basis for the most basic asset management information structure. The challenge is to identify those pieces of information that are necessary to the development of an asset management plan. More advanced structures require a wider range and higher level of information. These more advanced models may be developed in future WERF Strategic Asset Management research.

6.13 Asset Management Information Requirements Step by Step

Table 6-1 identifies the relevant minimum information required to execute the 10-step process to build an asset management plan. The items contained in the "Output Information" column result from applying basic asset management techniques to the "Associated Input Information" column. These techniques are hyperlinked to relevant sections of SIMPLE and can be explored in more detail by clicking to the relevant link. They are also integrated into the U.S. EPA Asset Management training, "The Fundamentals of Asset Management", also available in SIMPLE and on the U.S. EPA website (http://www.epa.gov/owm/assetmanage/assets_training.htm).

Table 6-1. Information Requirements for Each Step of Asset Management.

Step	Associated Input Information	Output Information
1. Develop asset registry	Asset ID, parent ID, child ID Responsible organizational unit Asset Type or Class Location Year of installation Original acquisition cost Manufacturer, make, model, serial number Images (photos, videos)	None
2. Assess performance, failure modes	Condition rating Reliability rating Performance rating Maintenance history Refurbishment history Maximum design capacity Actual utilization	None
3. Determine residual life	Engineer's or manufacturer's estimated design life	Estimated remaining physical useful life Dominant primary failure mode Imminent primary failure mode
4. Determine life cycle and replacement costs	Full acquisition cost Historic operations costs Historic maintenance costs Projected operations costs Projected maintenance costs Projected renewal costs (repair, refurbish, replace) Projected decommissioning costs	None
5. Set target levels of service	Target strategic levels of service metrics Actual strategic levels of service metrics Asset level target performance requirements Asset level actual performance requirements Projected utilization patterns	None
6. Determine business risk	Probability of failure rating Consequence of failure rating Redundancy/risk mitigation factor	Business risk exposure ("criticality") score
7. Optimize O&M investment	Is failure predictable for this asset? Is failure preventable for this asset?	None
8. Optimize capital investment	Number of times this asset can be refurbished Minimum condition level Maximum risk tolerance level	"At risk assets" list Asset issues list

Table 6-1. Asset Management Requirement by Step, continued.

Step	Associated Input Information	Output Information
9. Determine funding strategy	Funding availability Organizational financial condition	Projected annual capital investment requirement (with fund number) Projected annual operations investment requirement (with fund number) Projected annual maintenance investment requirement (with fund number) Required average annual investment
10. Build Asset Management Plan		Asset Management Plan or State of Assets Report: Strategy statements Projected cost patterns Funding strategies

6.14 A More Encompassing Asset Management Information Model

What might a more encompassing Asset Management information and system functionality model look like? How might information flows be mapped so that Asset Management investment strategies can not only be effectively and efficiently developed, but be constantly updated and refined as changes in both external and internal factors occur? In Appendix B, Figure B-1 depicts a more sophisticated Asset Management information schematic, one that a moderate to advanced level of Asset Management practice might deploy. It is presented solely for contextual purposes to illustrate to where a simpler start-up effort can systematically evolve as asset management information is converted to decision relevant knowledge.

APPENDIX A

Catalog of Existing Decision Support Tools for Asset Management

Table A-1. Catalog of Existing Decision Support Tools for Asset Management

Source	Tool or Document Name	Reference	Asset Management Area
ASTM	Standard practice for least cost (life cycle) analysis of concrete culvert, storm sewer, and sanitary sewer systems and other buried conduits. (D)	http://www.astm.org/DATABASE.CART/HISTORICAL/C1131-95R00.htm	Replacement Costing
ASTM	Standard practice for life cycle cost analysis of corrugated metal pipe used for culverts, storm sewers, and other buried conduits (D).	http://www.astm.org/Standards/A930.htm	Replacement Costing
ASTM	Standard practice for measuring life cycle cost analysis of plastic pipe used in culverts, storm sewers and other buried conduits (D).	http://www.astm.org/Standards/F1675.htm	Replacement Costing
AUS Dept of Sustainability	Fair Value Asset Valuation (D)	http://www.aasb.gov.au/admin/file/content105/c9/ACCED181_06-09.pdf	Asset Valuation
Austroads	Economic Evaluation of Road Investment Proposals (D) Numerous publications	http://www.onlinepublications.austroads.com.au	Asset Valuation
Austroads	Valuation of Road Infrastructure in Australia/New Zealand (D) AP-144/00	http://www.onlinepublications.austroads.com.au/script/Details.asp?DocN=AR0000024_0904	Asset Valuation
Austroads	Valuing the Environment and Other Externalities (D) AP-R229/03	http://www.onlinepublications.austroads.com.au/script/Details.asp?DocN=AR0000046_0904	Asset Valuation
Austroads	Guidelines for development of Levels of Service Framework based on Community Consultation (D) AP-R289/06	http://www.onlinepublications.austroads.com.au/script/Details.asp?DocN=AS472782535440	Level of Service
Boise State EFC	CAP Finance (T)	http://efc.boisestate.edu/efc/LinkClick.aspx?link=90&tabid=58	Funding Requirements
Brisbane City Council	What are the Tools for Economic Appraisal (D)	http://www.brisbane.qld.gov.au/	Capital Investment Optimization, Maintenance

			Investment Optimization
CA Office of Information Security and Privacy	Information Security Risk Assessment (T)	http://www.cio.ca.gov/OIS/Government/risk/toolkit.asp	Business Risk
Commonwealth Scientific and Industrial Reasearch Org. (CSIRO)	Long Term Performance Prediction for PE Pipes, (D)	http://www.waterresearchfoundation.org/research/TopicsAndProjects/execSum/2975.aspx- 8.6KB - Project Profiles / Order Reports	Condition Assessment
	Long Term Performance Prediction for PVC Pipes, (D)	http://www.waterresearchfoundation.org/research/TopicsAndProjects/execSum/2879.aspx- 8.2KB - Project Profiles / Order Reports	Condition Assessment
Edmonton City of	Standardized System to Assess the Condition of Infrastructure in Edmonton (T)	http://www.edmonton.ca/city_government/documents/InfraPlan/Infra2002Report-FINAL.pdf	Condition Assessment, Business Risk, Major Failure Modes
FEMA	Mitigation BCA (Benefit Cost Analysis Toolkit (T)	http://www.fema.gov/government/grant/bca.shtm#1	Condition Assessment, Replacement Costing, Business Risk, Capital Investment Optimization,
INGENIUM - NAMS	International Infrastructure Management Manual (D)	http://www.nams.org.nz/pages/72/international-infrastructure-management-manual.htm	Asset Registry, Condition Assessment, Residual Life, Asset Valuation, Replacement Costing, Level of Service, Business Risk, Capital Investment Optimization, Maintenance Investment Optimization, Funding Requirements, AM Planning, Training
LMI Gov. Consulting	Performance Risk Assessment	http://www.lmi.org/index.aspx	Business Risk
	PHAST TM		Asset Registry

	Prime-Choice Decision Support Method (trademarked)		Capital Investment Optimization, Maintenance Investment Optimization
NACWA	Asset Based Vulnerability Checklist for Wastewater Utilities (T)	http://www.nacwa.org/images/stories/public/2002avcheck.pdf	Condition Assessment, Business Risk
	Continual improvement in utility management: A framework for integration(D)	http://www.nacwa.org/images/stories/public/2004-02CIGuide.pdf?phpMyAdmin=PM8UfvMmlxx8xqqtLrO9xEOmDg0	AM Planning
	Emergency Response Plan Guidelines for Wastewater Systems (D)	http://www.nacwa.org/images/stories/public/2004-10WastewaterERP.pdf?phpMyAdmin=PM8UfvMmlxx8xqqtLrO9xEOmDg0	Condition Assessment, Business Risk
	Vulnerability Self-Assessment Tool (VSAT) (T)	http://www.vsatusers.net/product.html	Condition Assessment, Business Risk
NZ Dept of Treasury	Cost Benefit Analysis Primer (D)	http://www.treasury.govt.nz/publications/guidance/costbenefitanalysis/primer	Capital Investment Optimization, Maintenance Investment Optimization
	Valuation Guidelines for Plant, Property and Equipment (D)	http://www.treasury.govt.nz/publications/guidance/valuation/property/val-guidance-property.pdf	Asset Valuation, Replacement Costing
United Kingdom Water Industry Research (UKWIR)	Long Term Performance Prediction for PE Pipes, (D)	http://www.waterresearchfoundation.org/research/TopicsAndProjects/execSum/2975.aspx- 8.6KB - Project Profiles / Order Reports	Condition Assessment
	Long Term Performance Prediction for PVC Pipes, (D)	http://www.waterresearchfoundation.org/research/TopicsAndProjects/execSum/2879.aspx- 8.2KB - Project Profiles / Order Reports	Condition Assessment
USACE	ECONPACK (T)	http://www.hnd.usace.army.mil/paxspt/econ/download.aspx	Asset Valuation, Replacement Costing, Capital Investment Optimization, Maintenance Investment Optimization

US DOE	Building Life Cycle Costing (BLCC) Program (T)	http://www1.eere.energy.gov/femp/information/download_blcc.html	Asset Valuation, Replacement Costing, Funding Requirements
USDOT,FHWA	The Life Cycle Cost Analysis Primer (T)	http://www.fhwa.dot.gov/infrastructure/asstmgmt/lcca.cfm	Replacement Costing
USEPA	Asset Management - A Handbook for Small Water Systems	http://www.epa.gov/ogwdw/smallsystems/pdfs/guide_smallsystems_asset_mgmnt.pdf	Asset Registry
	Capital Asset Plan and Business Case Summary	http://www.epa.gov	Funding Requirements
	CUPSS – Check Up Program for Small Systems	http://www.epa.gov/ogwdw/cupss	Asset Registry, Condition Assessment, Residual Life, Asset Valuation, Level of Service, Business Risk, AM Planning, Training
	Sources of Technical & Financial Assistance for Small Drinking Water Systems	http://www.epa.gov/ogwdw000/arsenic/pdfs/funding/tfa_sdws.pdf	Funding Requirements
	Strategic Planning: A Handbook for Small Water Systems	http://www.epa.gov/ogwdw/smallsystems/pdfs/guide_smallsystems_stratplan.pdf	Asset Registry, AM Planning
	Taking Stock of your Water System - A Simple Inventory for Very Small Drinking Water Systems	http://www.epa.gov/OGWDW/smallsystems/pdfs/final_asset_inventory_for_small_systems.pdf	Asset Registry
	TEAMS – Total Electronic Asset Management System	http://www.mcet.org/Technical/environment/teamsAM.html	Asset Registry, Condition Assessment, Residual Life, Asset Valuation, Capital Investment Optimization, Maintenance Investment Optimization, Funding Requirements,Training
	U.S. EPA Asset Management Workshop Training Material	http://www.epa.gov/OW-OWM.html/assetmanage/assets_training.htm	Training

	SIMPLE (T)	http://www.werf.org/am/custom/simple_gateway.cfm	Asset Registry, Condition Assessment, Residual Life, Asset Valuation, Replacement Costing, Level of Service, Business Risk, Capital Investment Optimization, Maintenance Investment Optimization, Funding Requirements, Major Failure Modes, Training
WERF	Condition Assessment Strategies and Protocols for Water and Wastewater Utility Assets	http://www.werf.org/AM/Template.cfm?Section=Search&Template=/CustomSource/Research/ResearchProfile.cfm&ReportId=03-CTS-20CO&ID=03-CTS-20CO	Condition Assessment
	Remaining Asset Life: A State of the Art Review	http://www.werf.org/AM/Template.cfm?Section=Search&Template=/CustomSource/Research/ResearchProfile.cfm&ReportId=SAM1R06d&ID=SAM1R06d	Condition Assessment, Residual Life
	Risk Management Tool	http://www.werf.org/am/custom/simple_gateway.cfm	Condition Assessment, Residual Life, Replacement Costing, Level of Service, Business Risk, Capital Investment Optimization, Maintenance Investment Optimization, Funding Requirements, Major Failure Modes,

	SCRAPS - Sewer cataloguing, retrieval, and prioritization system	http://www.werf.org/AM/Template.cfm?Section=Search&Template=/CustomSource/Research/ResearchProfile.cfm&ReportId=97-CTS-7&ID=97-CTS-7	Condition Assessment, Residual Life, Replacement Costing, Level of Service, Business Risk, Capital Investment Optimization, Maintenance Investment Optimization, Major Failure Modes,
WaterNZ	New Zealand Infrastructure Grading Guidelines – 1999 (D)	http://www.waternz.org.nz/bookshop.html#manuals	Condition Assessment
	New Zealand Pipe Inspection Manual -2006 3rd Ed (D)	http://www.waternz.org.nz/bookshop.html#manuals	Condition Assessment
	Assessment of Non-Uniform Corrosion in Copper Piping (D)	http://www.waternz.org.nz/bookshop.html#manuals	Condition Assessment
	Workshop on Condition Assessment Inspection Devices for Water Transmission Mains (D)	http://www.waternz.org.nz/bookshop.html#manuals	Condition Assessment
WRc	GRASP (T)	solutions@wrcplc.co.uk	Condition Assessment
	ProRIsk (T)	solutions@wrcplc.co.uk	Major Failure Modes
	SEESAW (T)	solutions@wrcplc.co.uk	Capital Investment Optimization, Maintenance Investment Optimization
	TR 61 (T)	solutions@wrcplc.co.uk	Capital Investment Optimization, Maintenance Investment Optimization
	Waterfowl (T)	solutions@wrcplc.co.uk	Replacement Costing
WaterRF	Long Term Performance Prediction for PE Pipes, (D)	http://www.waterresearchfoundation.org/research/TopicsAndProjects/execSum/2975.aspx- 8.6KB - Project Profiles / Order Reports	Condition Assessment

	Long Term Performance Prediction for PVC Pipes, (D)	http://www.waterresearchfoundation.org/research/TopicsAndProjects/execSum/2879.aspx- 8.2KB - Project Profiles / Order Reports	Condition Assessment
	Failure of Pre stressed Concrete Cylindrical Pipes	http://www.waterresearchfoundation.org/research/TopicsAndProjects/execSum/4034.aspx- 7.7KB - Project Profiles / Order Reports	Condition Assessment

Legend:
D = document
T = tool

APPENDIX B

ADVANCED ASSET MANAGEMENT INFORMATION SYSTEM LIFE CYCLE DATA FLOW EXAMPLE

Table B-1. Advanced Asset Management Information System Life Cycle Data Flow Example.

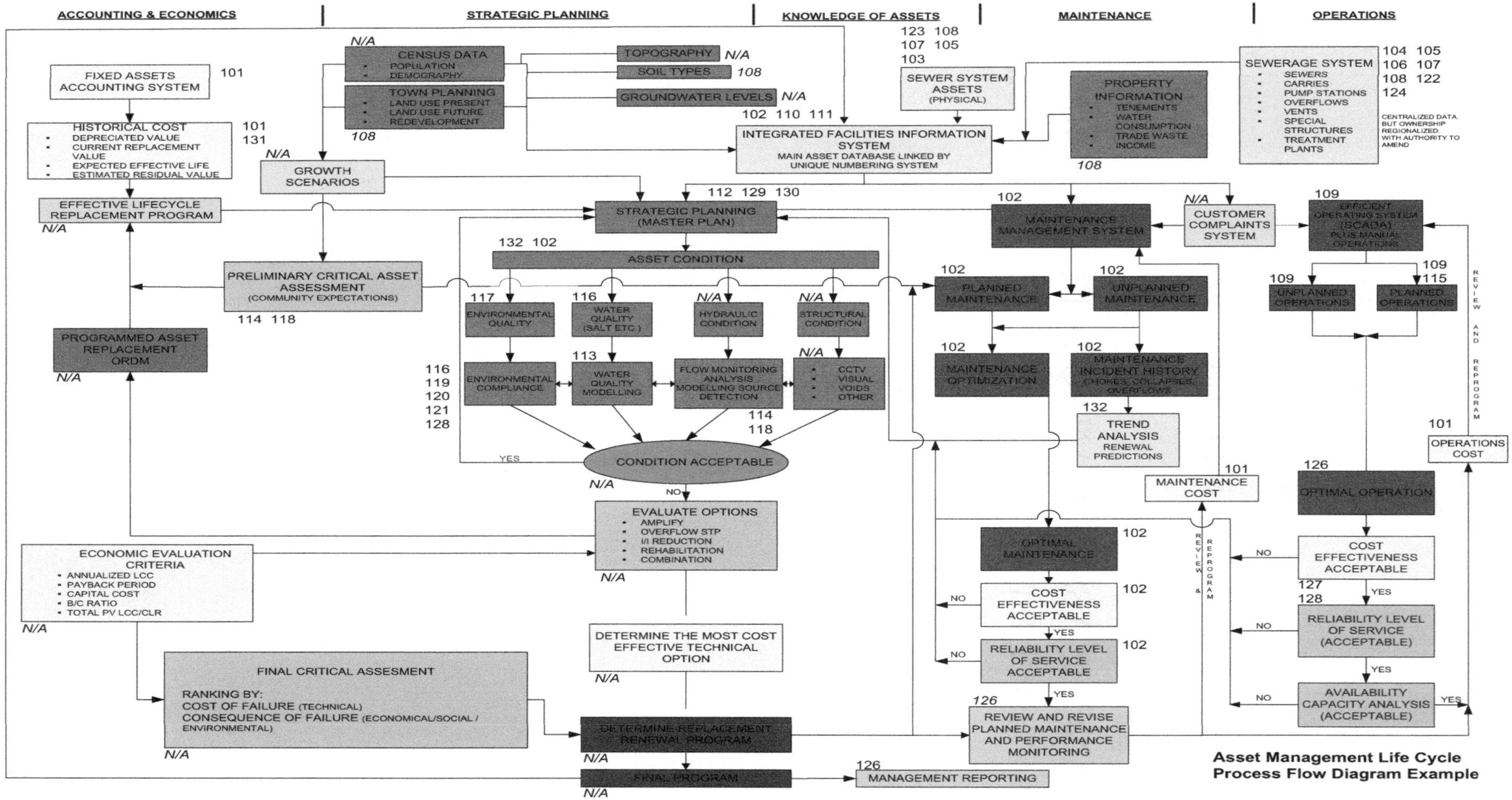

WASTEWATER UTILITY

Alabama
Montgomery Water Works & Sanitary Sewer Board

Alaska
Anchorage Water & Wastewater Utility

Arizona
Avondale, City of
Glendale, City of, Utilities Department
Mesa, City of
Peoria, City of
Phoenix Water Services Dept.
Pima County Wastewater Management
Safford, City of
Tempe, City of

Arkansas
Little Rock Wastewater Utility

California
Central Contra Costa Sanitary District
Corona, City of
Crestline Sanitation District
Delta Diablo Sanitation District
Dublin San Ramon Services District
East Bay Dischargers Authority
East Bay Municipal Utility District
El Dorado Irrigation District
Fairfield-Suisun Sewer District
Fresno Department of Public Utilities
Inland Empire Utilities Agency
Irvine Ranch Water District
Las Gallinas Valley Sanitary District
Las Virgenes Municipal Water District
Livermore, City of
Los Angeles, City of
Los Angeles County, Sanitation Districts of
Napa Sanitation District
Novato Sanitary District
Orange County Sanitation District
Palo Alto, City of
Riverside, City of
Sacramento Regional County Sanitation District
San Diego Metropolitan Wastewater Department, City of
San Francisco, City & County of
San Jose, City of
Santa Barbara, City of
Santa Cruz, City of
Santa Rosa, City of
South Bayside System Authority
South Coast Water District
South Orange County Wastewater Authority
South Tahoe Public Utility District
Stege Sanitary District
Sunnyvale, City of
Union Sanitary District
West Valley Sanitation District

Colorado
Aurora, City of
Boulder, City of
Greeley, City of
Littleton/Englewood Water Pollution Control Plant
Metro Wastewater Reclamation District, Denver

Connecticut
Greater New Haven WPCA
Stamford, City of

District of Columbia
District of Columbia Water & Sewer Authority

Florida
Broward, County of
Fort Lauderdale, City of
Jacksonville Electric Authority (JEA)
Miami-Dade Water & Sewer Authority
Orange County Utilities Department
Pinellas, County of
Reedy Creek Improvement District
Seminole County Environmental Services
St. Petersburg, City of
Tallahassee, City of
Toho Water Authority
West Palm Beach, City of

Georgia
Atlanta Department of Watershed Management
Augusta, City of
Clayton County Water Authority
Cobb County Water System
Columbus Water Works
Fulton County
Gwinnett County Department of Public Utilities
Savannah, City of

Hawaii
Honolulu, City & County of

Idaho
Boise, City of

Illinois
Decatur, Sanitary District of
Greater Peoria Sanitary District
Kankakee River Metropolitan Agency
Metropolitan Water Reclamation District of Greater Chicago
Wheaton Sanitary District

Indiana
Jeffersonville, City of

Iowa
Ames, City of
Cedar Rapids Wastewater Facility
Des Moines, City of
Iowa City

Kansas
Johnson County Wastewater
Unified Government of Wyandotte County/ Kansas City, City of

Kentucky
Louisville & Jefferson County Metropolitan Sewer District
Sanitation District No. 1

Louisiana
Sewerage & Water Board of New Orleans

Maine
Bangor, City of
Portland Water District

Maryland
Anne Arundel County Bureau of Utility Operations
Howard County Bureau of Utilities
Washington Suburban Sanitary Commission

Massachusetts
Boston Water & Sewer Commission
Massachusetts Water Resources Authority (MWRA)
Upper Blackstone Water Pollution Abatement District

Michigan
Ann Arbor, City of
Detroit, City of
Holland Board of Public Works
Saginaw, City of
Wayne County Department of Environment
Wyoming, City of

Minnesota
Rochester, City of
Western Lake Superior Sanitary District

Missouri
Independence, City of
Kansas City Missouri Water Services Department
Little Blue Valley Sewer District
Metropolitan St. Louis Sewer District

Nebraska
Lincoln Wastewater & Solid Waste System

Nevada
Henderson, City of
Las Vegas, City of
Reno, City of

New Jersey
Bergen County Utilities Authority
Ocean County Utilities Authority

New York
New York City Department of Environmental Protection

North Carolina
Charlotte/Mecklenburg Utilities
Durham, City of
Metropolitan Sewerage District of Buncombe County
Orange Water & Sewer Authority
University of North Carolina, Chapel Hill

Ohio
Akron, City of
Butler County Department of Environmental Services
Columbus, City of
Metropolitan Sewer District of Greater Cincinnati
Montgomery, County of
Northeast Ohio Regional Sewer District
Summit, County of

Oklahoma
Oklahoma City Water & Wastewater Utility Department
Tulsa, City of

Oregon
Albany, City of
Clean Water Services
Eugene, City of
Gresham, City of
Portland, City of Bureau of Environmental Services
Lake Oswego, City of
Oak Lodge Sanitary District
Water Environment Services

Pennsylvania
Hemlock Municipal Sewer Cooperative (HMSC)
Philadelphia, City of
University Area Joint Authority

South Carolina
Charleston Water System
Mount Pleasant Waterworks & Sewer Commission
Spartanburg Water

Tennessee
Cleveland Utilities
Murfreesboro Water & Sewer Department
Nashville Metro Water Services

Texas
Austin, City of
Dallas Water Utilities
Denton, City of
El Paso Water Utilities

Fort Worth, City of
Houston, City of
San Antonio Water System
Trinity River Authority

Utah
Salt Lake City Corporation

Virginia
Alexandria Sanitation Authority
Arlington, County of
Fairfax County
Hampton Roads Sanitation District
Hanover, County of
Henrico, County of
Hopewell Regional Wastewater Treatment Facility
Loudoun Water
Lynchburg Regional Wastewater Treatment Plant
Prince William County Service Authority
Richmond, City of
Rivanna Water & Sewer Authority

Washington
Everett, City of
King County Department of Natural Resources
Seattle Public Utilities
Sunnyside, Port of
Yakima, City of

Wisconsin
Green Bay Metro Sewerage District
Kenosha Water Utility
Madison Metropolitan Sewerage District
Milwaukee Metropolitan Sewerage District
Racine, City of
Sheboygan Regional Wastewater Treatment
Wausau Water Works

Water Services Association of Australia
ACTEW Corporation
Barwon Water
Central Highlands Water
City West Water
Coliban Water Corporation
Cradle Mountain Water
Gippsland Water
Gladstone Area Water Board
Gold Coast Water
Gosford City Council
Hunter Water Corporation
Logan Water
Melbourne Water
Onstream
Power & Water Corporation
SEQ Water
South Australia Water Corporation
South East Water Limited
Sunshine Coast Water
Sydney Catchment Authority
Sydney Water
Wannon Regional Water Corporation
Watercare Services Limited (NZ)
Water Corporation
Water Distribution Brisbane City Council
Western Water
Yarra Valley Water

Canada
Edmonton, City of/Edmonton Waste Management Centre of Excellence
Lethbridge, City of
Regina, City of, Saskatchewan
Toronto, City of, Ontario
Winnipeg, City of, Manitoba

STORMWATER UTILITY

California
Fresno Metropolitan Flood Control District
Los Angeles, City of, Department of Public Works
Monterey, City of
San Francisco, City & County of
Santa Rosa, City of
Sunnyvale, City of

Colorado
Aurora, City of
Boulder, City of

Florida
Orlando, City of

Iowa
Cedar Rapids Wastewater Facility
Des Moines, City of

Kansas
Lenexa, City of
Overland Park, City of

Kentucky
Louisville & Jefferson County Metropolitan Sewer District

Maine
Portland Water District

North Carolina
Charlotte, City of, Stormwater Services

Pennsylvania
Philadelphia, City of

Tennessee
Chattanooga Stormwater Management

Texas
Harris County Flood Control District, Texas

Washington
Bellevue Utilities Department
Seattle Public Utilities

STATE

Connecticut Department of Environmental Protection
Kansas Department of Health & Environment
New England Interstate Water Pollution Control Commission (NEIWPCC)
Ohio Environmental Protection Agency
Ohio River Valley Sanitation Commission
Urban Drainage & Flood Control District, CO

CORPORATE

ADS LLC
Advanced Data Mining International
AECOM
Alan Plummer & Associates
Alpine Technology Inc.
Aqua-Aerobic Systems Inc.
Aquateam–Norwegian Water Technology Centre A/S
ARCADIS
Associated Engineering
Bernardin Lochmueller & Associates
Black & Veatch
Blue Water Technologies, Inc.
Brown & Caldwell
Burgess & Niple, Ltd.
Burns & McDonnell
CABE Associates Inc.
The Cadmus Group
Camp Dresser & McKee Inc.
Carollo Engineers Inc.
Carpenter Environmental Associates Inc.
CET Engineering Services
CH2M HILL
CRA Infrastructure & Engineering
CONTECH Stormwater Solutions
D&B/Guarino Engineers, LLC
Damon S. Williams Associates, LLC
Ecovation
EMA Inc.
Environmental Operating Solutions, Inc.
Environ International Corporation
Fay, Spofford, & Thorndike Inc.
Freese & Nichols, Inc.
ftn Associates Inc.
Gannett Fleming Inc.
Garden & Associates, Ltd.
Geosyntec Consultants
GHD, Inc.
Greeley and Hansen LLC
Hazen & Sawyer, P.C.
HDR Engineering Inc.
HNTB Corporation
Hydromantis Inc.
HydroQual Inc.
Infilco Degremont Inc.
Jason Consultants LLC Inc.
Jordan, Jones, & Goulding Inc.
KCI Technologies Inc.
Kelly & Weaver, P.C.
Kennedy/Jenks Consultants
Larry Walker Associates
LimnoTech Inc.
Lombardo Associates, Inc.
The Low Impact Development Center Inc.
Malcolm Pirnie Inc.
Material Matters, Inc.
McKim & Creed
MWH
NTL Alaska, Inc.
O'Brien & Gere Engineers Inc.
Odor & Corrosion Technology Consultants Inc.
Parametrix Inc.
Parsons
Post, Buckley, Schuh & Jernigan
Praxair, Inc.
RMC Water & Environment
Ross & Associates Ltd.
SAIC
Siemens Water Technologies
The Soap & Detergent Association
Smith & Loveless, Inc.
Southeast Environmental Engineering, LLC
Stearns & Wheler, LLC
Stone Environmental Inc.
Stratus Consulting Inc.
Synagro Technologies Inc.
Tetra Tech Inc.
Trojan Technologies Inc.
Trussell Technologies, Inc.
URS Corporation
Wallingford Software
Westin Engineering Inc.
Wright Water Engineers
Zoeller Pump Company

INDUSTRY

American Electric Power
American Water
Anglian Water Services, Ltd.
Chevron Energy Technology
The Coca-Cola Company
Dow Chemical Company
DuPont Company
Eastman Chemical Company
Eli Lilly & Company
InsinkErator
Johnson & Johnson
Merck & Company Inc.
Procter & Gamble Company
Suez Environnment
United Utilities North West (UUNW)
United Water Services LLC
Veolia Water North America

Note: List as of 11/10/09

IWA Publishing's authorised EU representative for General Product
Safety Regulations is Diane D'Arras, 15 rue Duret, 75116 Paris,
France, e-mail: safety@iwap.co.uk.

Printed and bound by CPI Group (UK) Ltd, Croydon, CR0 4YY
17/09/2026
02218808-0010